PRACTICAL AMPLIFIER DIAGRAMS

By
Jack Robin, Sound Engineer
And
Chester E. Lipman, Member
Of Institute of Radio Engineers
Amateur Station W6CDZ

AF887783

Reprinted in 2000 by Audio Amateur Press

Distributed by KCK Media Corp.

P R A C T I C A L   A M P L I F I E R · D I A G R A M S

TABLE OF CONTENTS

Part I

SCHEMATIC DRAWINGS AND EXPLANATORY NOTES — Page

TABLE OF CONTENTS (cont.)

AMPLIFIER MANUAL

INTRODUCTION

In this electronic age, one of the most commonly used devices is the amplifier. It has made possible magnification of minute indications of voltage, current and power into levels high enough to perform many useful tasks.

It is the purpose of this manual to present a series of amplifiers designed to cover the audio frequencies, frequencies that affect the human ear, and those that cover the entire range of sound. The average listener is most sensitive to sound from about 100 cycles, a low pitch, to 7000 cycles (a high pitch). In some instances, one can hear sound frequencies as low as 20 cycles and perhaps as high as 20,000 cycles. An audio frequency (A.F.) amplifier, then, operates within this band of frequencies, either entirely or in part, depending upon the purpose for which it was designed.

The majority of (A.F.) amplifiers described in this manual are for purposes of amplifying the usual range of recorded sound, broadcast and voice frequencies, or from 50 to 12,000 cycles. There are some instances of amplifiers described to operate up to 20,000 cycles, and these are designated as High Fidelity amplifiers. All amplifiers shown are of standard design and typical of their type. No attempt has been made to specify any but standard parts in the diagrams. Such parts are generally procurable in a number of reputable brands from any reliable radio supply house or radio jobber.

Each schematic diagram carries a listing of parts necessary to build the amplifier so that substitutions can be made if the exact part specified cannot be obtained. In substituting parts it is necessary that they be within tolerance so that the completed amplifier will operate properly. Unless otherwise indicated in the particular schematic, the following electrical tolerances are acceptable when substituting:

Capacity of electrolytic condensers-------------- 50%
Capacity of paper condensers--------------------- 30%
Capacity of mica condensers---------------------- 10%
Resistance of fixed resistors-------------------- 20%
Resistance of potentiometers--------------------- 30%
Voltage insulation of condensers-------- minimum specified
Wattage rating of resistors------------- minimum specified
Current rating of chokes---------------- minimum specified
Wattage rating of transformers---------- minimum specified

It is assumed the prospective constructor has sufficient background to read and interpret schematic diagrams. No physical parts layout is specified in each amplifier, but stress is rather placed upon the complete schematic. This is done since constructors seldomly build their equipment alike--particularily when many items of the same electrical specifications have entirely different physical shapes.

The main consideration in the location of parts on the chassis is that the input circuits be located as far as possible from the output circuits and power supply equipment.

In order to simplify and standardize the arrangement of the diagrams, a ground symbol (⏚) is used to show common connections. Likewise, to lessen the confusion of filament wires appearing on the diagrams, an arrow with a letter (⟶ x) shows continuity between other identical symbols.

For those who have had little or no experience in building amplifiers, it is advisable to build the more simple types of one, two or three tubes, making certain that the completed amplifier is functioning properly before attempting to construct a more complicated type. A section entitled "Servicing Your Amplifier," located in the back of this manual, may be of assistance in obtaining the desired results from your amplifier.

To you more experienced constructors, it is hoped that the following pages will provide information which many of you desire.

The Authors

1 TUBE 1 WATT AC/DC AMPLIFIER

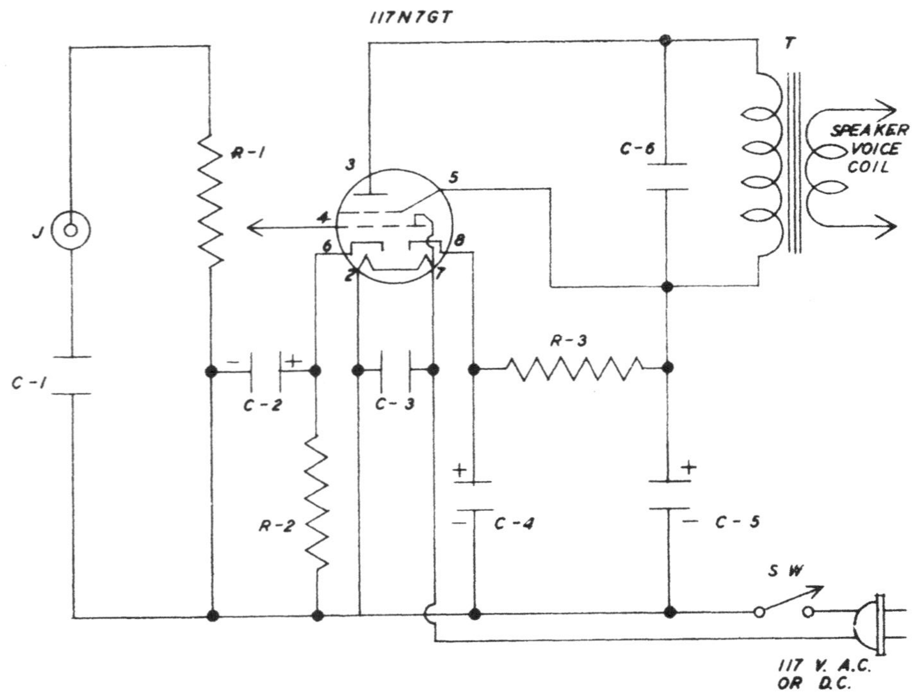

1 TUBE AC/DC WIRELESS PHONO. OSC.

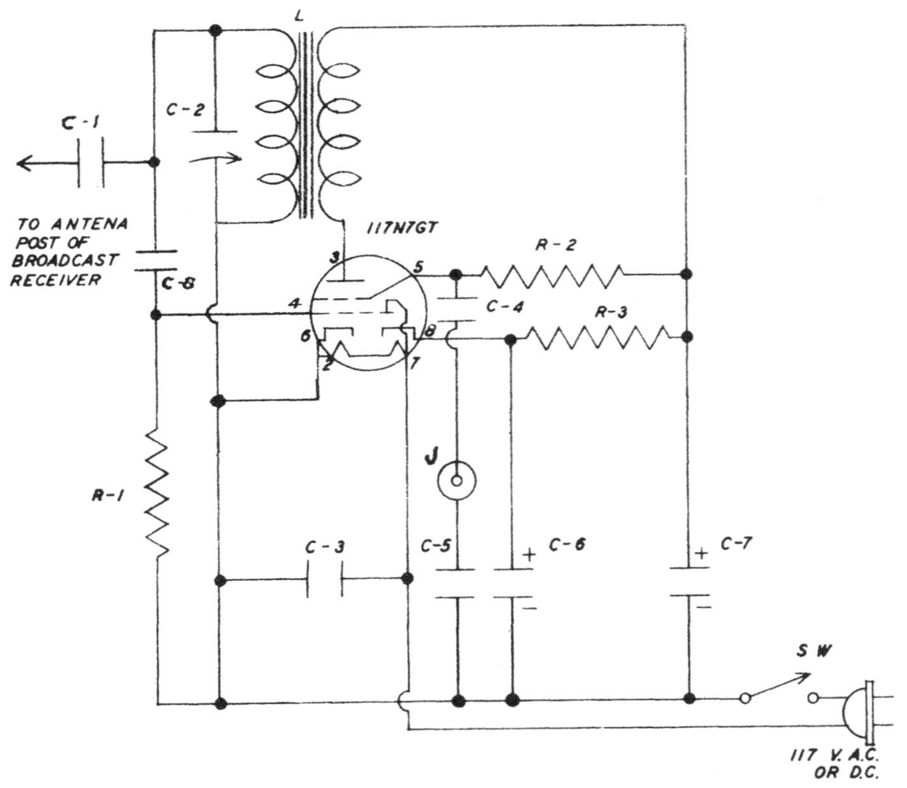

1 TUBE 1 WATT AC/DC AMPLIFIER

This amplifier, utilizing a dual power output-rectifier tube, is suitable for low volume from a phonograph pickup having at least two volts output. Either a dynamic or permanent magnet speaker may be used with equal results, provided directions are followed regarding filter choke L. The usual operation of an amplifier of this type is from a 115 volt A.C. source. If D.C. voltage is available it may be necessary to reverse the power plug if the amplifier will not operate.

Because of the few parts used, this amplifier can be built very small and compact, making it ideal to mount in a small phonograph case.

Parts List

C-1	.05 mfd 400v paper cond.	R-1	1 meg ohm volume control
C-2	10 mfd 25v elec. cond.	R-2	100 ohm 1 watt res.
C-3	.02 mfd 400v paper cond.	R-3	1000 ohm 1 watt res.
C-4	40 mfd 150v elec. cond.	J	Phonograph jack
C-5	40 mfd 150v elec. cond.	SW	SPST switch
C-6	.005 mfd 600v paper cond.	T	3000 ohms to voice coil

---

1 TUBE AC/DC "WIRELESS" PHONO. OSC.

The 117N7GT dual tube is again in a special type amplifier which is, in effect, a small broadcast transmitter. When connected to a broadcast receiver by means of a single wire to the antenna post, phonograph records can be played through the receiver. To operate, adjust the receiver tuning control to a clear frequency around 1500 KC, set the receiver volume control to normal, start a record through the receiver.

Coil L need not be of the shielded variety unless desired. It is a standard broadcast replacement type 456 KC oscillator coil. If, after completing the wireless oscillator, the records are not being reproduced in the receiver, try reversing the connections to one winding of coil L.

Very little trouble should be experienced in constructing this unit, and when once adjusted, should not require further attention for a long while. Caution: Do not run a separate ground lead to the oscillator as it is already grounded to the power line.

Parts List

C-1	50 mm mica cond.	R-1	40,000 ohm ½ watt res.
C-2	150 to 450mm padder cond.	R-2	3 meg ohm ½ watt res.
C-3	.02 mfd 400v paper cond.	R-3	3000 ohm 2 watt res.
C-4	.02 mfd 400v paper cond.	L	Broadcast osc. coil
C-5	.02 mfd 400v paper cond.	J	Phonograph jack
C-6	20 mfd 150v elec. cond.	SW	SPST switch
C-7	20 mfd 150v elec. cond.	Sockets:	1 octal
C-8	.0001 mica cond.		

2 TUBE 1½ WATT AC/DC AMPLIFIER

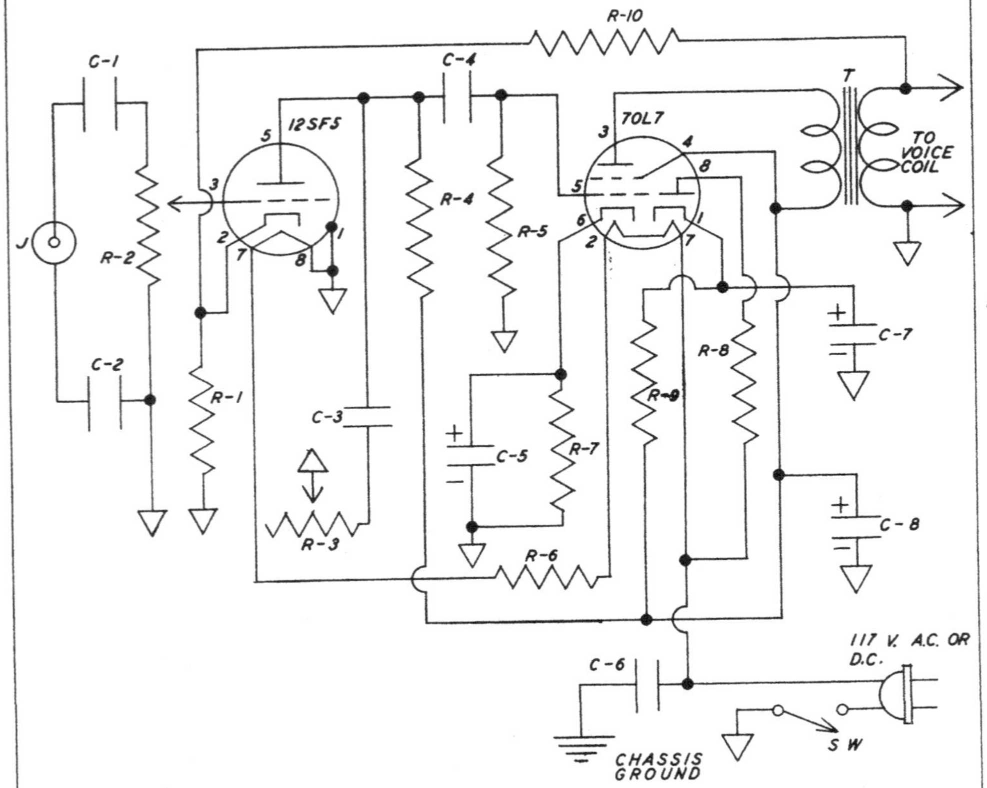

NOTE - ▽ INDICATES COMMON CONNECTIONS ISOLATED FROM CHASSIS

2 TUBE CONTACT MICROPHONE AMP

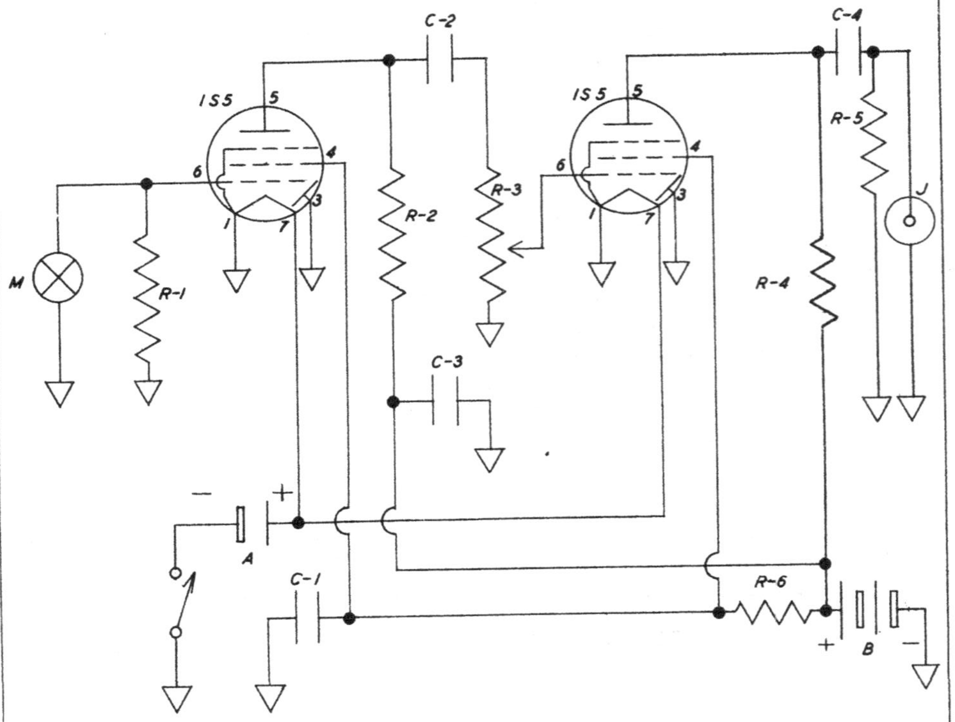

2 TUBE 1½ WATT AC/DC AMPLIFIER

Because of its greater gain, this amplifier is capable of fair volume with any type crystal phonograph pickup or radio tuner. It delivers up to 1½ watts output, which is about all an average 3" speaker can handly. The amplifier is of the AC/DC type and features an inverse feedback network to improve fidelity, and a tone control. As is true of all amplifiers shown, the power switch can either be a seperate toggle, rotary or push type switch, or it may be attached to the volume or tone control shafts as part of the operation of that particular control, according to the desires of the constructor.

There should be no particular problem in getting the completed amplifier to operate properly. If the amplifier has a tendancy to "howl" or oscillate, reverse the connections to the secondary, i.e., voice coil, side of the output transformer. This amplifier will make a very compact unit, especially since no filter choke or power transformer is used.

Parts List

C-1	.01 mfd 400 v. paper cond.	R-2	500,000 ohm vol. control
C-2	.05 mfd 400 v. paper cond.	R-3	100,000 " tone control
C-3	.02 mfd 400 v. paper cond.	R-4	250,000 " ½ watt res.
C-4	.02 mfd 600 v. paper cond.	R-5	500,000 " ½ watt res.
C-5	10 mfd 25 v. elec. cond.	R-6	235 ohm 10 watt res.
C-6	.02 mfd 400 v. paper cond.	R-7	150 " 1 watt res.
C-7	20 mfd 150 v. elec. cond.	R-8	100 " 1 " "
C-8	20 mfd 150 v. elec. cond.	R-9	2000 ohm 2 watt res.
R-1	2500 ohm ½ watt res.	R-10	25,000 ohm ½ watt res.
		T	Output trans: 2000 ohm to voice coil

---

2 TUBE CONTACT MICROPHONE AMPLIFIER

This amplifier is a high gain unit suitable for increasing to headphone level sounds of very small amplitude. The contact microphone is a crystal or magnetic unit attached to an instrument sounding board, table or any other surface resonant to sound. Just to name two uses for an instrument of this type, it could be used by a physician to detect malfunctions of the body or by police as a detecting device.

Battery power is used on this device for quietness and minature type tubes for compactness, making the whole portable. The microphone should not be placed over 10 feet from the amplifier and shielded microphone wire used if best results are to be obtained.

Parts List

C-1	.05 mfd 400v paper cond.	R-2	1 meg. ohm ½w res.	"A"	1½ v. dry cell
C-2	.01 mfd 400v paper cond.	R-3	2 meg. " vol. con.	"B"	67½ v. min. battery
C-3	.05 mfd 400v paper cond.	R-4	1 meg. " ½w res.	M	Crystal or mag. contact mic.
C-4	.01 mfd 400v paper cond.	R-5	500,000 " ½w res.		
R-1	10 meg. ohm ½ watt res.	R-6	1 meg. " ½w res.	J	Headphone jack
					Sockets-2 min. 7 pin

2 TUBE 4 WATT AC BOOSTER AMP.

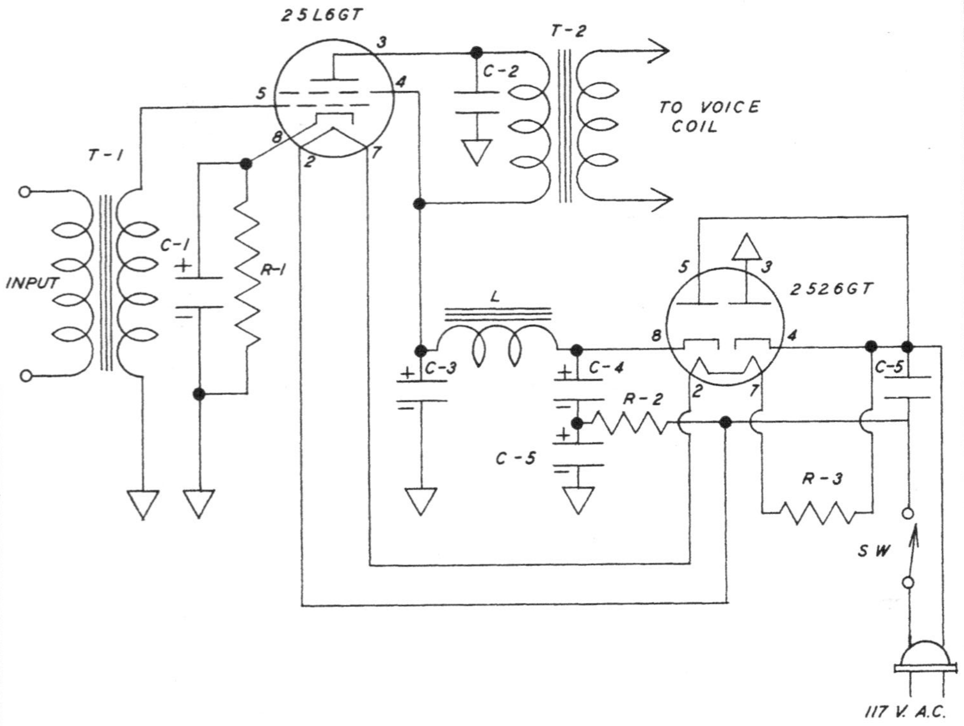

2 TUBE 5 WATT A.C. BOOSTER AMP.

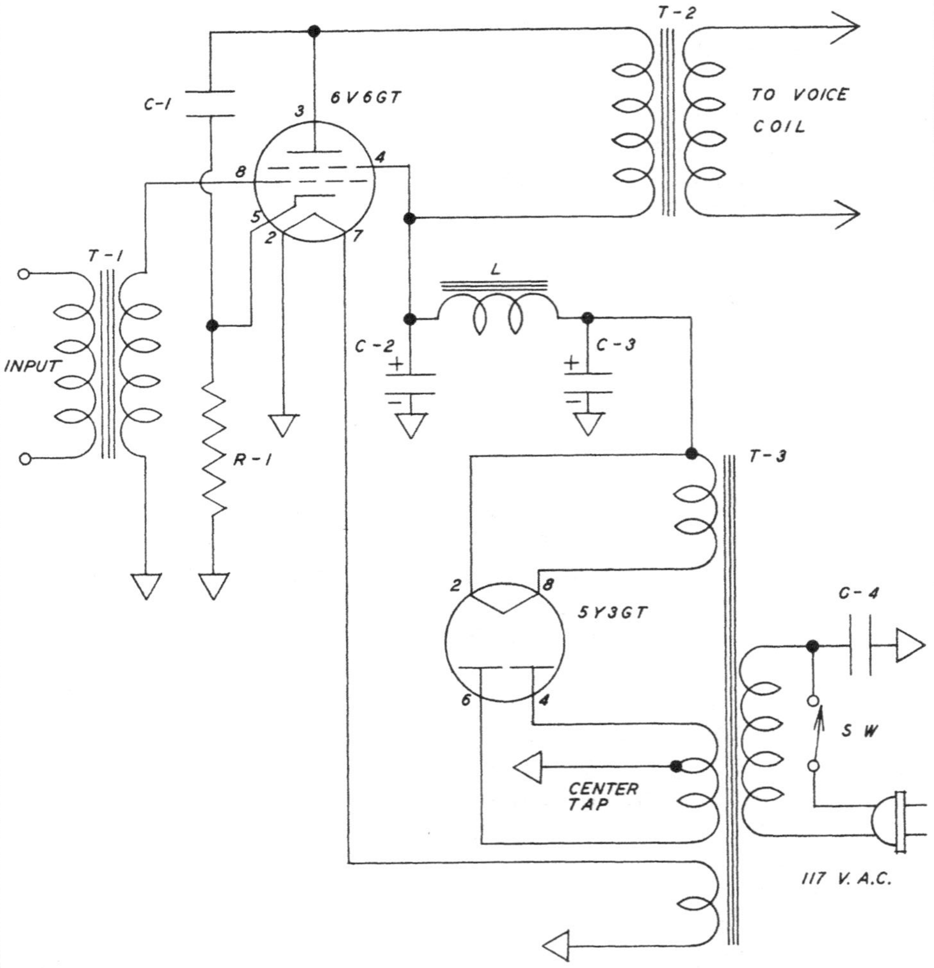

2 TUBE 4 WATT AC BOOSTER AMPLIFIER "TRANSFORMERLESS"

The purpose of this low gain, one stage amplifier, is to boost the output of a radio tuner or pre-amplifier to good room volume. Although this is a "transformerless" A.C. amplifier, using a single power tube, about 4 watts output can be obtained due to the voltage doubler rectifier circuit used. Any type of permanent magnet (PM) speaker can be used to the amplifier, provided the output transformer is properly matched. Likewise, the input transformer will depend upon the type of input used. If a radio tuner output is connected to the primary of T1, a 3:1 ratio audio transformer may be used; if a magnetic phonograph pickup is required, the primary of T1 must match this pickup unit.

Fair quality can be expected from the speaker with this circuit. In constructing the amplifier, be certain T1 is located as far as possible from the output transformer or filter choke to minimize inductive hum problems.

2 TUBE 5 WATT AC BOOSTER AMPLIFIER

This is the "power transformer" version of the previous amplifier. Slightly more power output is available due to higher plate voltages used, and somewhat better stability is realized due to the isolation of the amplifier and the A.C. power source. However, the overall gain is less due to degeneration applied to the 6V6Gt cathode circuit. This degeneration aids in improving the fidelity of the amplifier but at a sacrifice in its gain. As a result the amplifier requires a fairly high input level in order to realize its 5 watt output possibility. The overall fidelity will depend upon the quality of T1, T2, and the particular speaker used.

2 TUBE 4 WATT AC/DC BOOSTER AMPLIFIER
Parts List

C-1	10 mfd 25v elec. cond.
C-2	.005 mfd 600 v. paper cond.
C-3	20 mfd 250 v. elec. cond.
C-4	20 mfd 250 v. elec. cond.
C-5	20 mfd 250 v. elec. cond.
C-6	.02 mfd 400 v. paper cond.
R-1	150 ohm 1 watt res.
R-2	50 ohm 1 watt res.
R-3	225 ohm 25w res. or line cord
T-1	Input transformer
T-2	Output transformer. 3000 ohm to v.c.
L	200 ohm 60ma filter choke
SW	spst switch
Sockets	2 octals

2 TUBE 5 WATT AC BOOSTER AMPLIFIER
Parts List

C-1	.01 mfd 600 v. paper cond.
C-2	20 mfd 450 v. elec. cond.
C-3	20 mfd 450 v. elec. cond.
C-4	.05 mfd 400 v. paper cond.
R-1	250 ohm 2 watt res.
L	300 ohm 70ma filter choke
SW	spst switch
T-1	Input transformer
T-2	Output trans. 8500 ohm to voice coil
T-3	Power trans: 300-0-300 @ 70ma 5 v @ 2a 6.3 v @ 1a
Sockets	2 octals

2 TUBE BATTERY INTERCOM. AMP.

"A"

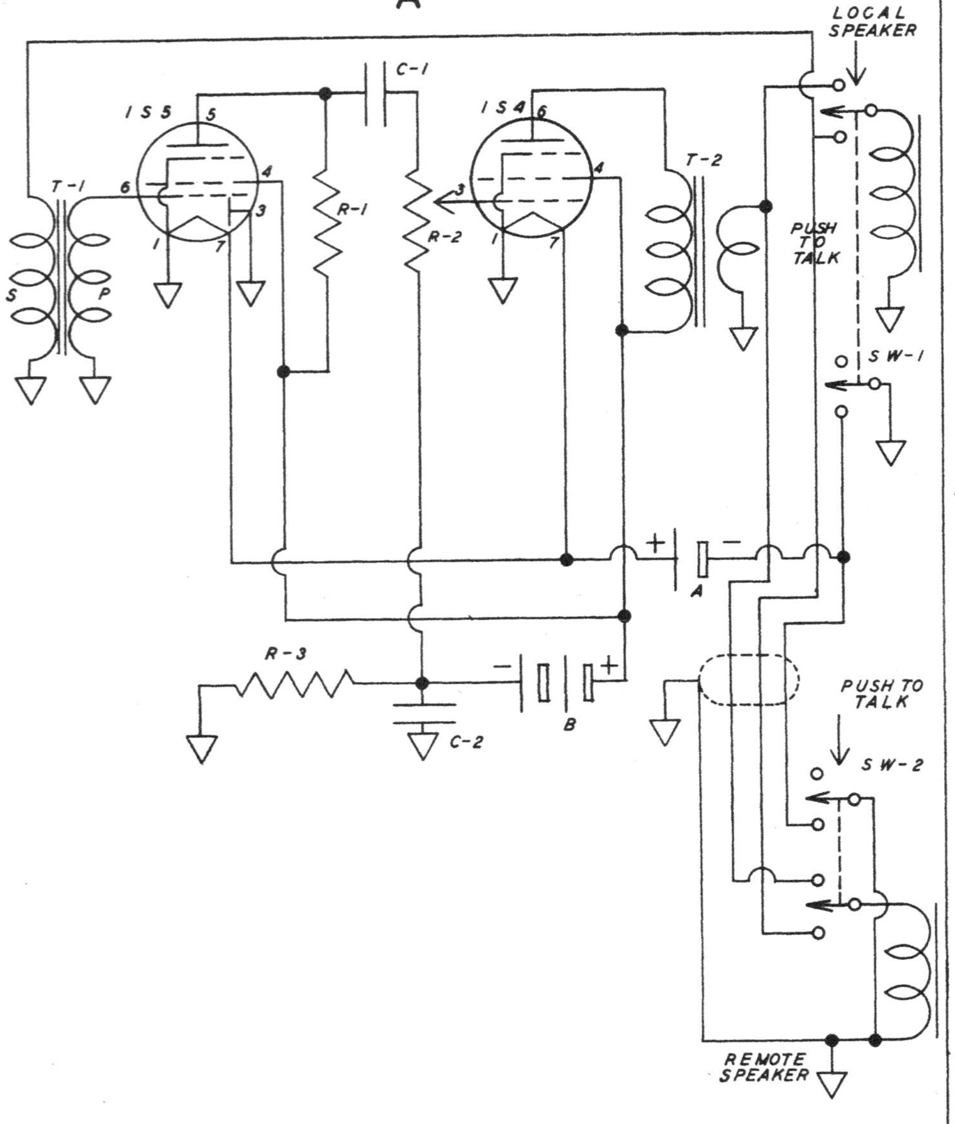

2 TUBE AC/DC INTERCOM. AMP.

"B"

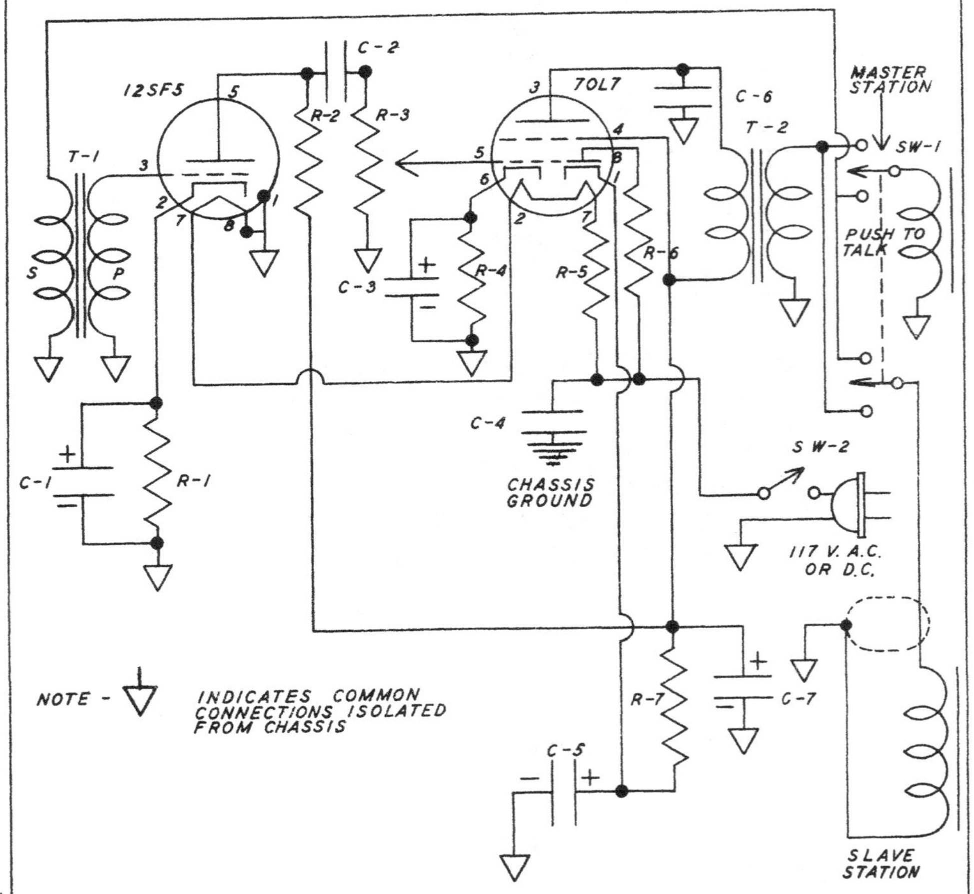

NOTE - INDICATES COMMON CONNECTIONS ISOLATED FROM CHASSIS

2 TUBE BATTERY INTERCOM AMPLIFIER AND 2 TUBE AC/DC INTERCOM AMPLIFIER

These two intercommunication type amplifiers are very similar, but for the power supply. "A" is designed for battery operation and is therefore instantly useable by pressing the push-to-talk switch on either speaker unit. This connects the voice coil to the amplifier properly and also turns on the tube filaments. A 4-wire cable is necessary on this system and for best results, is limited to 100 feet between remote speaker and amplifier. "B" is designed for AC/DC operation, therefore is characteristically a slow heater. It is best used as shown, a master (local) station having all control, and a slave (remote) station normally open to the master. A 2-wire cable is all that is necessary with this arrangement and up to 500 feet can be used without difficulty.

The speakers are 2" to 5" PM units and the push-to-talk switches are of the push button spring return types. In assembly, especially in amplifier "B", care should be taken in locating the input transformer T1. If hum persists regardless of its location, try reversing either primary or secondary winding of T1.

2 TUBE BATTERY INTERCOM AMPLIFIER
Parts List

C-1	.01 mfd 400 v. paper cond.
C-2	.1 mfd 400 v. paper cond.
R-1	250,000 ohm $\frac{1}{2}$w res.
R-2	1 meg. ohm vol. control
R-3	700 ohm 1 watt res.
T-1	Output trans. 25,000 ohm to v.c.
T-2	Output trans. 5000 ohm to v.c.
A	$1\frac{1}{2}$ v. dry cell, flashlight type
B	$67\frac{1}{2}$ v. minature battery
Speakers:	2" to 5" PM
Sockets:	2 minature 7 pin
sw-1	DPDT push to talk switch
sw-2	DPDT push to talk switch

2 TUBE AC/DC INTERCOM AMPLIFIER
Parts List

C-1	10 mfd 25 v. elec. cond.
C-2	.01 mfd 400 v. paper cond.
C-3	10 mfd 25v elec. cond.
C-4	.005 mfd 400 v. elec. cond.
C-5	40 mfd 150 v. elec. cond.
C-6	.01 mfd 600 v. paper cond.
C-7	40 mfd 150 v. elec. cond.
R-1	2500 ohm $\frac{1}{2}$ watt res.
R-2	250,000 ohm $\frac{1}{2}$ watt res.
R-3	1 meg ohm vol. control
R-4	150 ohm 1 watt res.
R-5	235 ohm 10 watt res.
R-6	100 ohm 1 watt res.
T-1	Output trans: 25,000 ohm to voice coil
T-2	Output trans: 2000 ohm to voice coil
R-7	2000 ohm 2 watt res.
sw-1	DPDT push to talk switch
sw-2	SPST switch
Speakers:	2" to 5" PM
Sockets:	2 octals

3 TUBE-2 WATT ECONOMY AC/DC AMPLIFIER

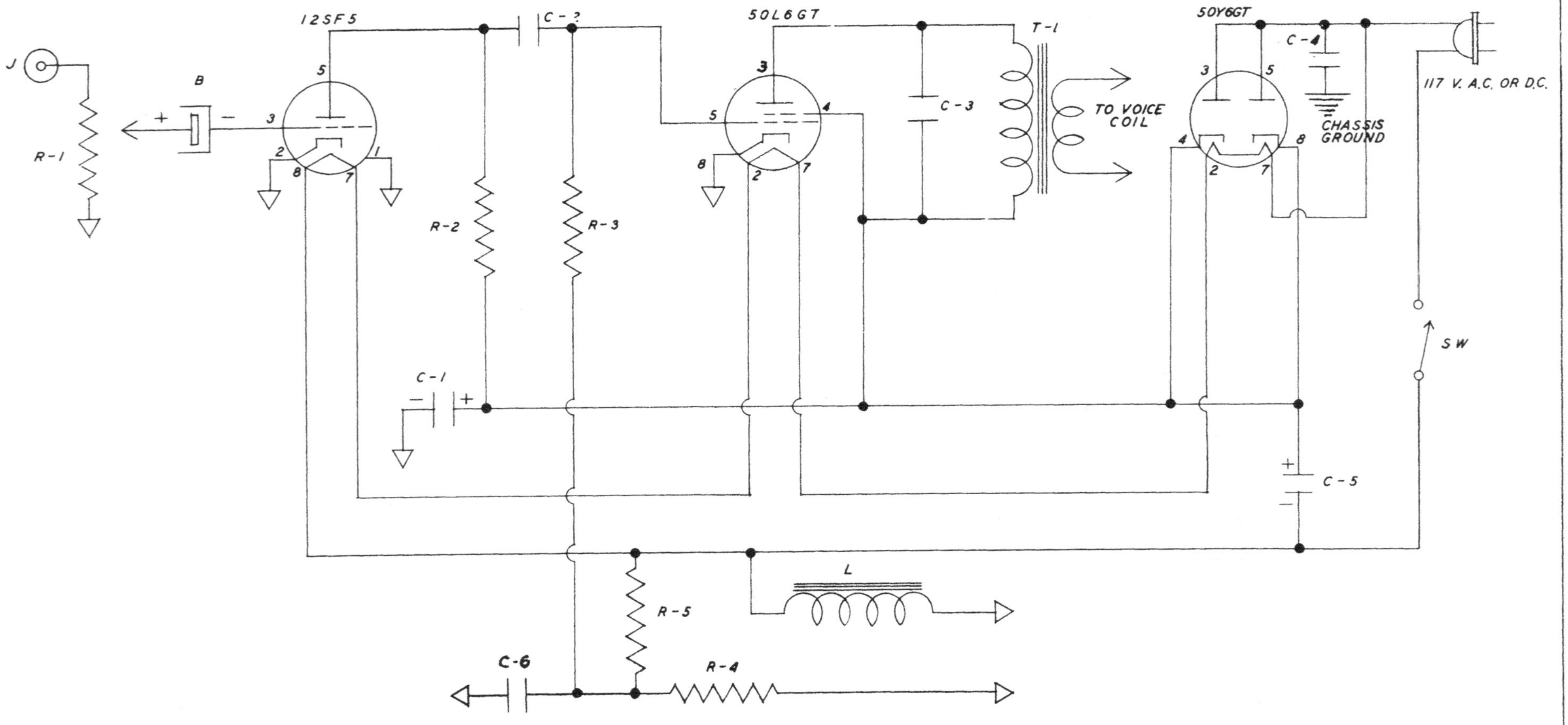

NOTE - ▽ INDICATES COMMON CONNECTIONS ISOLATED FROM CHASSIS

3 TUBE 2 WATT ECONOMY AC/DC AMPLIFIER

The feature of this AC/DC amplifier is its low cost due to the small number of parts used. It is quite suitable for phonograph reproduction with good gain capabilities from a less sensitive pickup. With the arrangement shown for biasing the two amplifier tubes, it is possible to ground the cathodes of the tubes directly, eliminating the need of cathode resistors and condensers. The Bias Cell provides essential grid bias voltage to the 12SF5 tube when connected in the proper manner--shell of bias cell to grid terminal. Bias for the 50L6GT tube is obtained from the divider network R-4 and R-5, connected across the speaker field. This method of biasing provides a slightly higher plate voltage to be effective on the power tube, consequently increasing its power output. It should be noted that the two filter condensers do not have their negative leads connected to the same point, so if it is intended to use a dual filter unit in the construction of this amplifier, one should not be obtained with a common negative lead.

Parts List

C-1	20 mfd 150 v. elec. cond.
C-2	.01 mfd 400 v. paper cond.
C-3	.01 mfd 600 v. paper cond.
C-4	.02 mfd 400 v. paper cond.
C-5	20 mfd 150 v. elec. cond.
R-1	1 meg ohm volume control
R-2	250,000 ohm $\frac{1}{2}$ watt res.
R-3	250,000 ohm $\frac{1}{2}$ watt res.
R-4	35,000 ohm $\frac{1}{2}$ watt res.
R-5	100,000 ohm $\frac{1}{2}$ watt res.
J	Phonograph jack
B	Bias cell $1\frac{1}{4}$ volt
L	450 ohm speaker field
T	Output trans: 2000 ohm to v.c.
sw	SPST switch on volume control
Sockets:	3 octals
C-6	.1 mfd. 200 v. paper cond.

3 TUBE 3 WATT DIRECT COUPLED AMPLIFIER

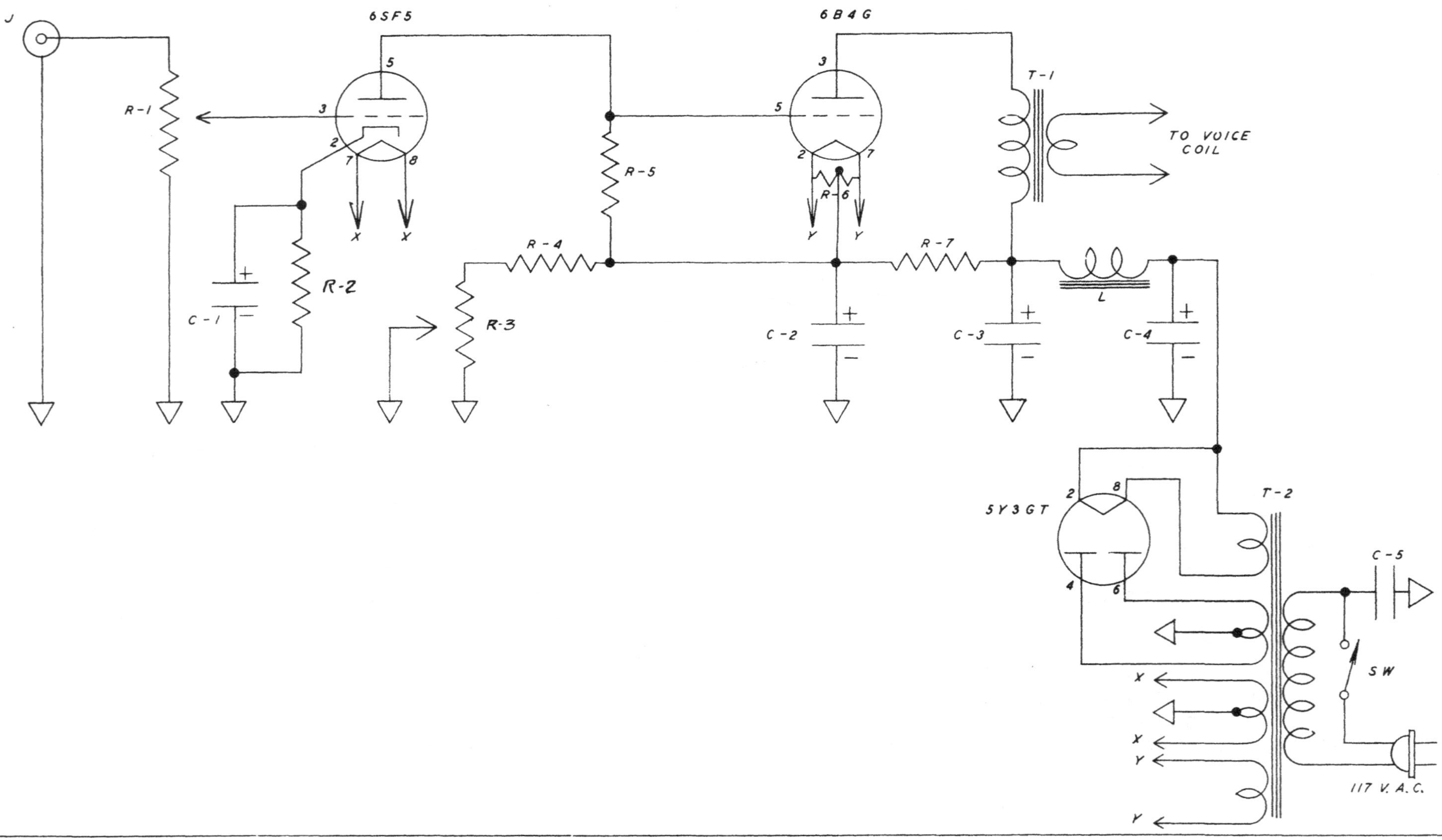

3 TUBE 3 WATT DIRECT COUPLED AMPLIFIER

The direct coupled amplifier is always a difficult type of amplifier to construct and keep in adjustment, but is presented here for the more ambitious constructor. As you know, the direct coupled circuit consists of the plate of an amplifier tube directly connected to the grid of the next amplifier tube, thereby allowing a complete transfer of signal variations with a minimum of frequency distortion. With the proper choice of tubes and transformers, a high fidelity amplifier can be constructed when properly balanced. This balancing adjustment consists of varying R-2, and the semi-adjustable resistor set to such a position that 45 volts can be measured across R-5 by means of a vacuum tube voltmeter. Three watts undistorted power output can easily be obtained with any phonograph pickup or radio tuner connected to the input.

Parts List

C-1	10 mfd 25 v. elec. cond.
C-2	20 mfd 250 v. elec. cond.
C-3	20 mfd 450 v. elec. cond.
C-4	8 mfd 600 v. elec. cond.
C-5	.05 mfd 400 v. paper cond.
R-1	500,000 ohm vol. control
R-2	3000 ohm ½ watt res.
R-3	5000 ohm 25 w semi-adj. res.
R-4	1000 ohm 10 watt res.
R-5	100,000 ohm 1 watt res.
R-6	20 ohm 5w center tap. res.
R-7	25,000 ohm 10 watt res.
T-1	Output trans: 2500 ohm to voice coil 5 watt
T-2	Power trans: 325-0-325 v @ 75ma; 5 v @ 2a; 6.3 v @ 1a; 6.3 v @ 1a
L	Filter choke 200 ohm 75ma
J	Input jack
SW	SPST switch
Sockets:	3 octals

3 TUBE 4 WATT A.C. AMPLIFIER

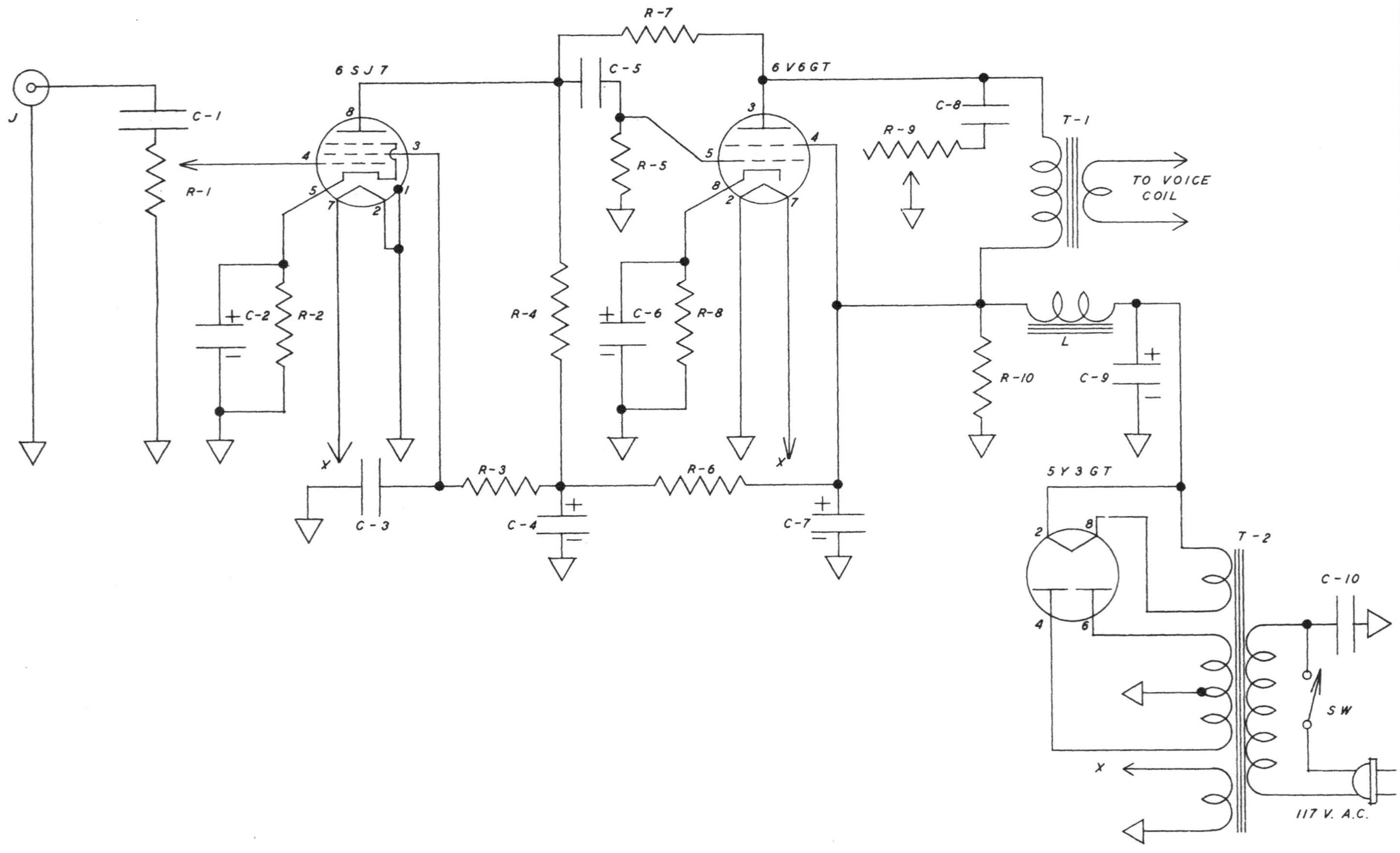

3 TUBE 4 WATT AC AMPLIFIER

For a small amplifier having good gain and tone, the reliable circuit shown is very satisfactory. It will operate well from either a phonograph pickup or radio tuner and deliver good room volume.

Very little difficult should be experienced in constructing this unit. Due to the use of the decoupling resistor, R-6, stability is greatly improved--resulting in better tone quality. A tone control is incorporated to provide a reduction of the higher audio frequencies when desired.

Parts List

C-1	.01 mfd 600 v. paper cond.
C-2	25 mfd 25 v. elec. cond.
C-3	.1 mfd 400 v. paper cond.
C-4	8 mfd 450 v. elec. cond.
C-5	.05 mfd 600 v. paper cond.
C-6	10 mfd 50 v. elec. cond.
C-7	16 mfd 450 v. elec. cond.
C-8	.05 mfd 600 v. paper cond.
C-9	16 mfd 450 v. elec. cond.
C-10	.1 mfd 400 v. paper cond.
R-1	500,000 ohm vol. control
R-2	2000 ohm ½ watt res.
R-3	1 meg ohm ½ watt res.
R-4	250,000 ohm ½ watt res.
R-5	500,000 ohm ½ watt res.
R-6	25,000 ohm 1 watt res.
R-7	1 meg ohm ½ watt res.
R-8	250 ohm 1 watt res.
R-9	100,000 ohm tone control
R-10	25,000 ohm 10 watt res.
T-1	Output trans: 5000 ohm to voice coil
T-2	Power trans: 300-0-300 v @ 60ma, 5 v @ 2a, 6.3 v @ 2a
L	Filter choke: 300 ohm 60ma
J	Input jack
SW	SPST switch on tone or volume control
Sockets:	3 octals

3 TUBE 4 WATT A.C. "TRANSFORMERLESS" AMPLIFIER

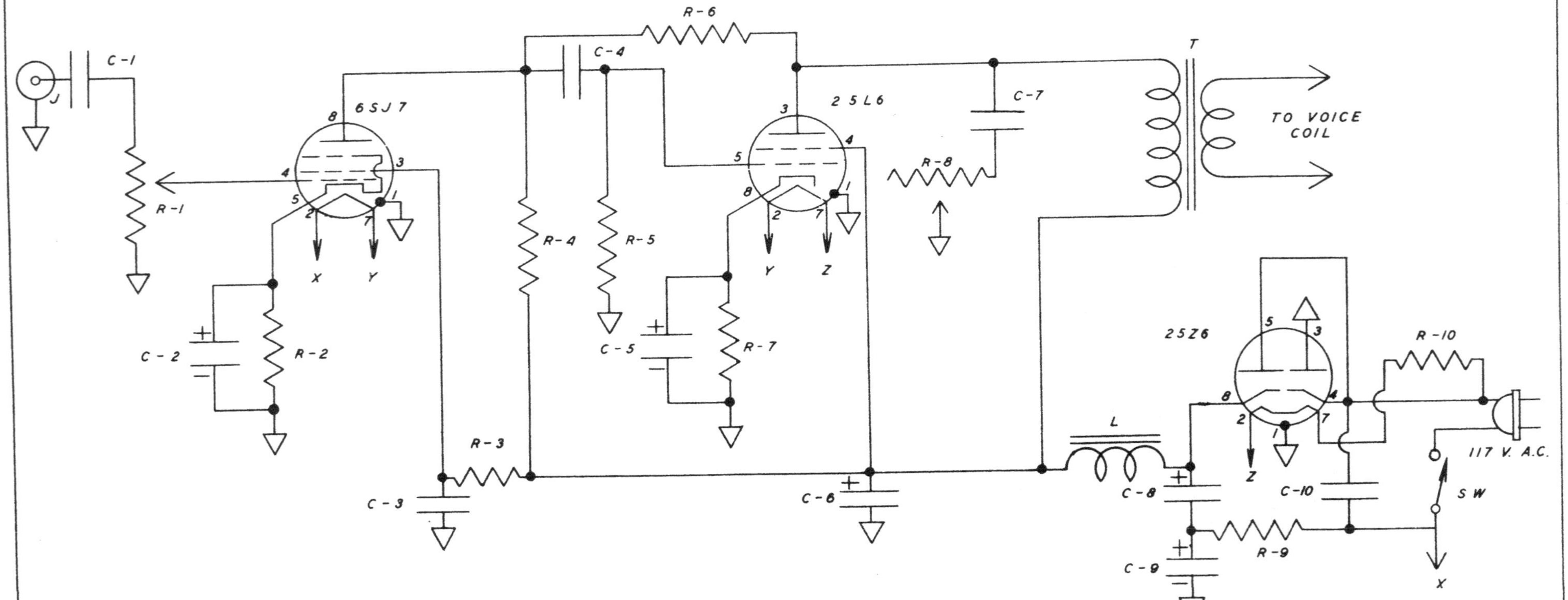

3 TUBE 4 WATT AC "TRANSFORMERLESS" AMPLIFIER

Although this amplifier circuit is very similar to the previous one, its chief difference lies in the power supply. It uses a voltage doubler rectifier circuit to furnish the higher voltages necessary to obtain the 4 watts output from a 25L6 tube. This is made possible without the aid of a power transformer, consequently the size and weight of the amplifier will be much less than the former. R-10 can be either a line cord resistor, ballast tube, or fixed resistor, depending upon the constructors choice.

Parts List

C-1	.01 mfd 600 v. paper cond.
C-2	10 mfd 25 v. elec. cond.
C-3	.05 mfd 400 v. paper cond.
C-4	.01 mfd 600 v. paper cond.
C-5	10 mfd 25 v. elec. cond.
C-6	20 mfd 250 v. elec. cond.
C-7	.05 mfd 600 v. paper cond.
C-8	20 mfd 250 v. elec. cond.
C-9	20 mfd 250 v. elec. cond.
C-10	.02 mfd 400 v. paper cond.
R-1	500,000 ohm volume control
R-2	2000 ohm $\frac{1}{2}$ watt res.
R-3	1 meg ohm $\frac{1}{2}$ watt res.
R-4	250,000 ohm $\frac{1}{2}$ watt res.
R-5	500,000 ohm $\frac{1}{2}$ watt res.
R-6	1.5 meg ohm $\frac{1}{2}$ watt res.
R-7	150 ohm 1 watt res.
R-8	100,000 ohm tone cont.
R-9	50 ohm $\frac{1}{2}$ watt res.
R-10	200 ohm line cord res.
T	Output trans: 3000 ohm to voice coil
L	Filter choke: 200 ohm 60ma
SW	SPST switch on tone or volume control
Sockets:	3 octals
J	Input jack

3 TUBE 8 WATT A.C. AMPLIFIER

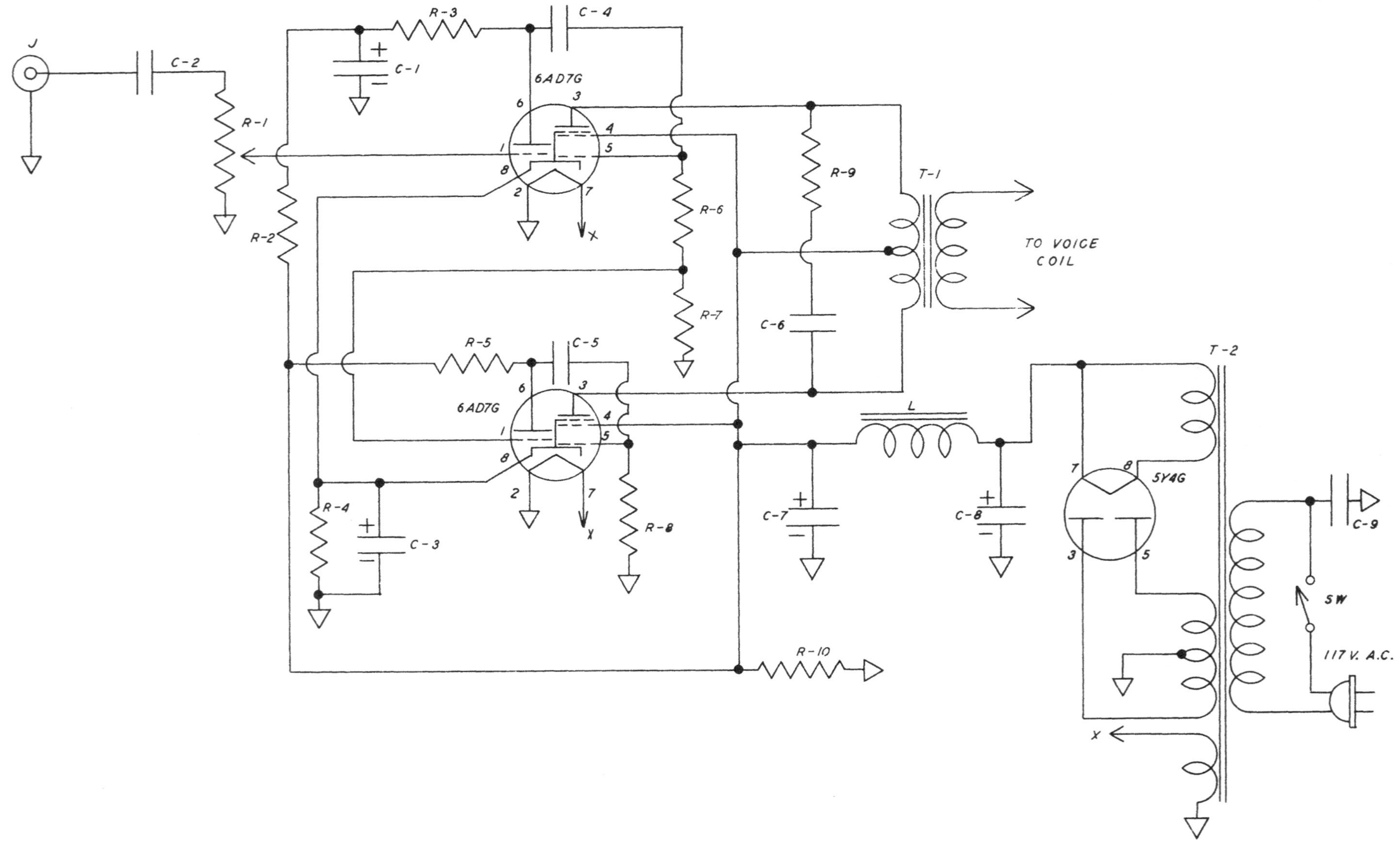

3 TUBE 8 WATT AC AMPLIFIER

This is a unique amplifier in that three tubes are doing the work of a normal five tube A.C. amplifier. The 6AD7G's are combination triode-power pentode tubes, and as connected, results in a two stage push-pull amplifier capable of about 8 watts output. Being in push-pull, the power tubes effectively reduce the even harmonic distortion, thereby considerably improving the amplifier's fidelity. The triode portion of one 6AD7G tube is a first stage amplifier, while the triode portion of the other 6AD7G is a phase inverter stage to provide push-pull operation to the pentode portions of the two tubes. A fairly strong input voltage can be fed into the amplifier from either a phonograph pickup or radio tuner.

Assembly of this amplifier is somewhat more involved than in the previous types, due to the greater number of parts used, therefore greater care must be exercised in its construction. Since two similar tubes are used, be careful that you do not confuse the wiring to the tube sockets.

Parts List

C-1	8 mfd 450 v. elec. cond.
C-2	.02 mfd 400 v. paper cond.
C-3	25 mfd 50 v. elec. cond.
C-4	.02 mfd 600 v. paper cond.
C-5	.02 mfd 600 v. paper cond.
C-6	.03 mfd 600 v. paper cond.
C-7	20 mfd 450 v. elec. cond.
C-8	20 mfd 450 v. elec. cond.
C-9	.05 mfd 400 v. paper cond.
R-1	500,000 ohm volume control
R-2	10,000 ohm 1 watt res.
R-3	50,000 ohm 1 watt res.
R-4	500 ohm 5 watt res.
R-5	50,000 ohm 1 watt res.
R-6	425,000 ohm 1 watt res.
R-7	75,000 ohm 1 watt res.
R-8	500,000 ohm 1 watt res.
R-9	20,000 ohm 1 watt res.
R-10	25,000 ohm 10 watt res.
T-1	Output trans: 14,000 ohm to v.c.
T-2	Power trans: 300-0-300 v @ 75ma; 5 v @ 2a; 6.3 v @ 2a
L	Filter choke: 250 ohm 75ma
J	Input Jack
sw	SPST switch on volume control
Sockets:	3 octals

3 TUBE QUICK HEATING A.C. PAGING AMPLIFIER

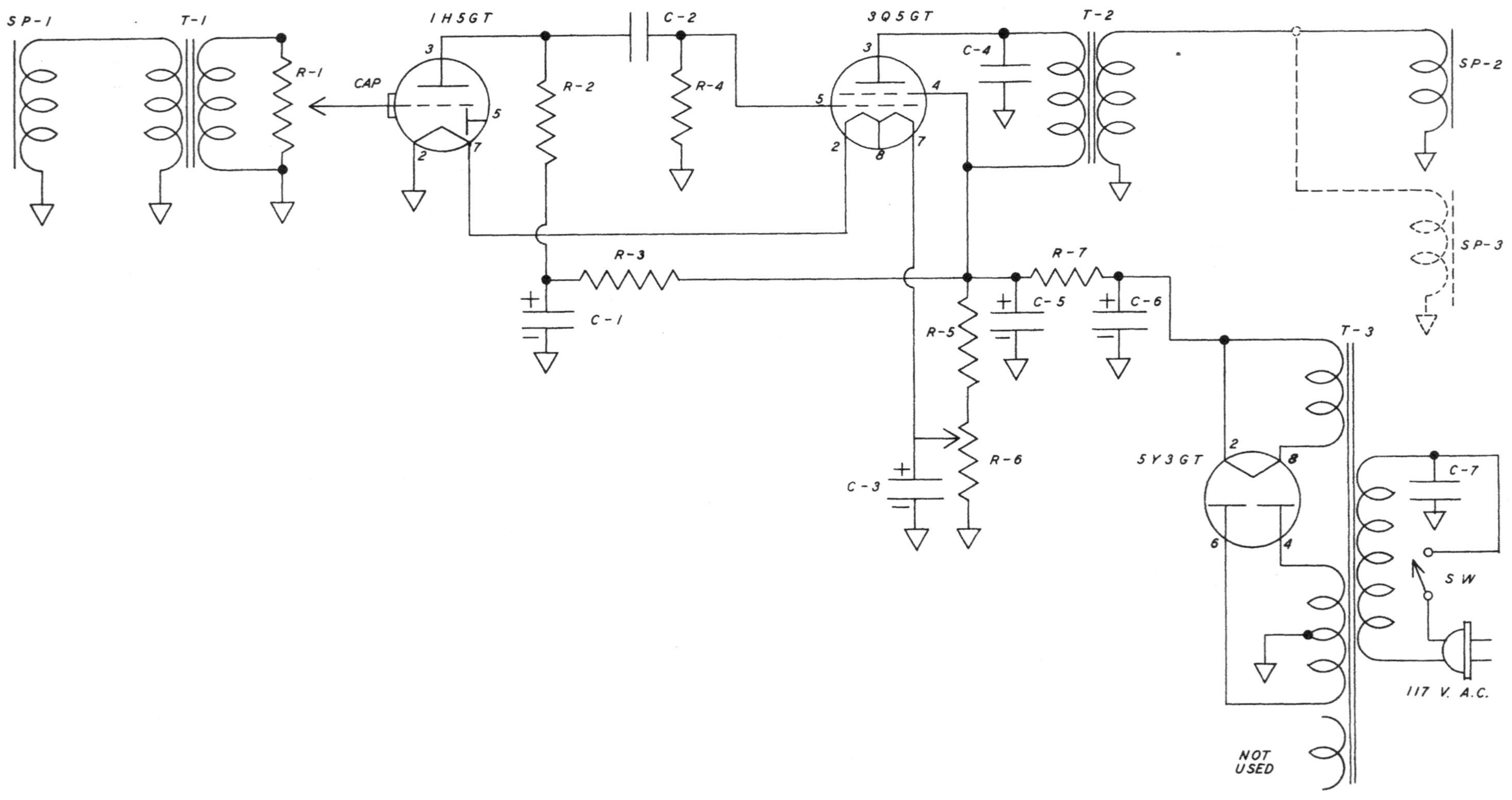

3 TUBE QUICK HEATING AC INTERCOM AMPLIFIER

The amplifier shown is an A.C. communication type which features battery type amplifier tubes for fast operation. This particular amplifier is convenient where only occasional operation is desired as a paging system. When ready to operate, the switch is first closed, turning on the amplifier. After approximately five seconds, it is permissable to talk into the system as it will then be in full operation. Any reasonable number of remote stations may be used in this system if care is taken in matching the output transformer to the speakers. Low volume is to be expected from the amplifier, but a volume control is incorporated in order to have complete control of the output.

In assembling the amplifier, carefully separate input transformer, T-1, from the other transformers to eliminate inductive hum pickup. In adjusting R-6 for correct filament voltage, start the tap from the ground end. With a DC voltmeter connected across C-3, move the tap on R-6 away from the grounded end until $4\frac{1}{2}$ volts is indicated on the meter.

Parts List

C-1	.1 mfd 400 v. paper cond.
C-2	.006 mfd 600 v. paper cond.
C-3	100 mfd 10 v. elec. cond.
C-4	.006 mfd 600 v. paper cond.
C-5	8 mfd 450 v. elec. cond.
C-6	40 mfd 450 v. elec. cond.
C-7	.05 mfd 400 v. paper cond.
R-1	1 meg. ohm vol. control
R-2	$\frac{1}{2}$ meg ohm $\frac{1}{2}$ watt res.
R-3	200,000 ohm $\frac{1}{2}$ watt res.
R-4	1 meg ohm $\frac{1}{2}$ watt res.
R-5	1500 ohm 20 watt res.
R-6	500 ohm 10w semi. adj. res.
R-7	3000 ohm 25 watt res.
sw	SPST switch--only momentary, normally off
sp-1	PM speaker or low impedance dynamic microphone
sp-2	5" PM remote speaker
T-1	Output trans; 25,000 ohm to v.c. connected in reverse
T-2	Output trans: 8000 ohm to v.c.
T-3	Power trans: 200-0-200 v @ 90ma; 5 v @ 2a
Sockets:	3 octals

3 TUBE 1½ WATT INTERCOM. AMPLIFIER

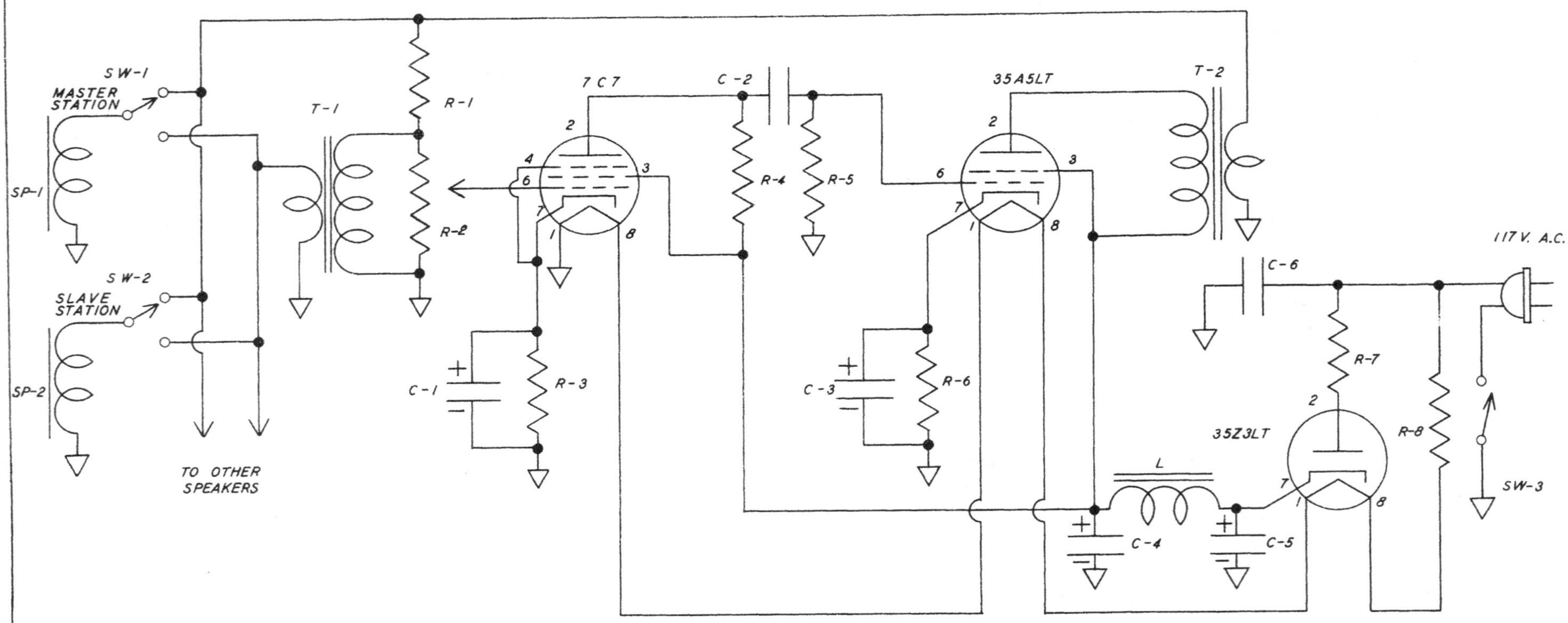

3 TUBE 1½ WATT AC/DC INTERCOM AMPLIFIER

Here is a reliable AC/DC intercommunication amplifier which is easy to build and simple to operate. It is designed around loctal type tubes and therefore can be made quite small and compact. Up to 5 or 6 remote stations can be used in the system, any one of which may talk to all the others by operation of sw-1, sw-2, etc. Inverse feedback is incorporated through R-1, making the amplifier quite stable.

Parts List

C-1	5 mfd 25 v. elec. cond.
C-2	.01 mfd 600 v. paper cond.
C-3	5 mfd 25 v. elec. cond.
C-4	20 mfd 150 v. elec. cond.
C-5	20 mfd 150 v. elec. cond.
C-6	.05 mfd 400 v. paper cond.
R-1	2 meg ohm ½ watt res.
R-2	500,000 ohm vol. control
R-3	1500 ohm ½ watt res.
R-4	250,000 ohm ½ watt res.
R-5	500,000 ohm ½watt res.
R-6	175 ohm 1 watt res.
R-7	50 ohm 1 watt res.
R-8	260 ohm 10 watt res.
L	Filter choke: 200 ohm @ 60ma
T-1	Output trans: 25,000 ohm to v.c. (connected in reverse)
T-2	Output trans: 2,500 ohm to v.c.
sp-1	3" to 6" speakers in amplifier
sp-2	3" to 6" PM speaker in sub-station
sw-1	SPDT push-to-talk switch
sw-2	SPDT push-to-talk switch
sw-3	SPST toggle switch
Sockets:	3 loctals

3 TUBE HEARING AID BATTERY AMPLIFIER

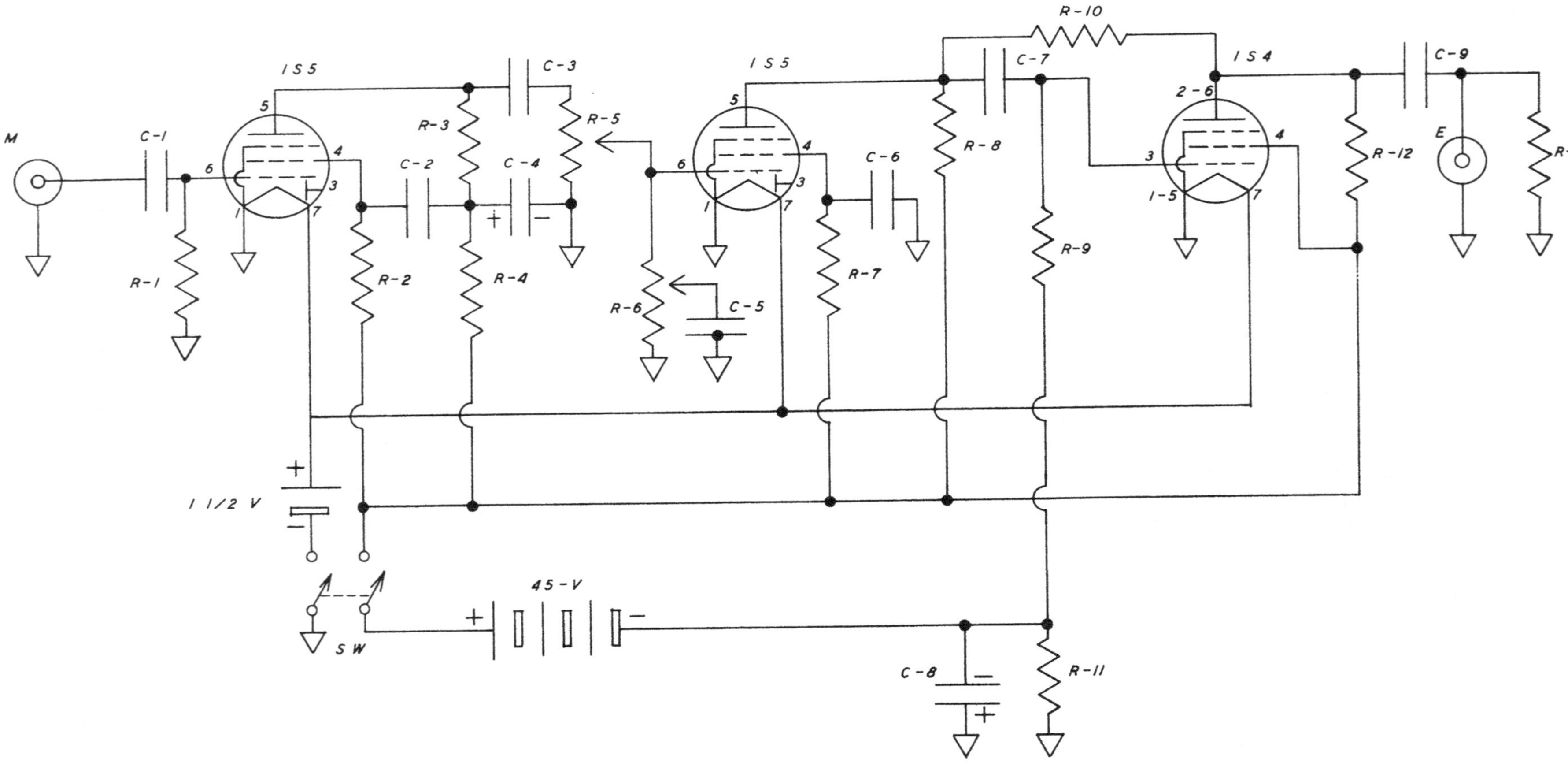

3 TUBE HEARING AID BATTERY AMPLIFIER

A high gain battery operated hearing aid amplifier is shown, consisting of minature type tubes for compactness and light weight. With a good microphone and earpiece, excellent results can be obtained. It is usually desireable to construct the amplifier so that it can be carried in a side pocket or be strapped to the waist of the user. The batteries are generally assembled in a separate container so that the amplifier bulk may be reduced. A volume and tone control is provided for adjusting the response of the amplifier to suit individual taste.

Parts List

C-1	.01 mfd 400 v. paper condenser
C-2	.02 mfd 400 v. paper condenser
C-3	.01 mfd 400 v. paper condenser
C-4	20 mfd 150 v. elec. condenser
C-5	.005 mfd 400 v. paper condenser
C-6	.02 mfd 400 v. paper condenser
C-7	.01 mfd 400 v. paper condenser
C-8	10 mfd 25 v. elec. condenser
C-9	.1 mfd 400 v. paper condenser
R-1	10 meg ohm ½ watt res.
R-2	3 meg ohm ½ watt res.
R-3	1 meg ohm ½ watt res.
R-4	20,000 ohm ½ watt res.
R-5	2 meg ohm vol. control
R-6	500 meg ohm tone control
R-7	3 meg ohm ½ watt res.
R-8	1 meg ohm ½ watt res.
R-9	2 meg ohm ½ watt res.
R-10	5 meg ohm ½ watt res.
R-11	1000 ohm 1 watt res.
R-12	8000 ohm 2 watt res.
R-13	1 meg ohm ½ watt res.
M	Lapel crystal mic.
E	Crystal earpiece
sw	DPST switch
Sockets:	3 minature
Batteries:	1½ v. "A"
	45 v. "B"

4 TUBE 5 WATT AC/DC AMPLIFIER

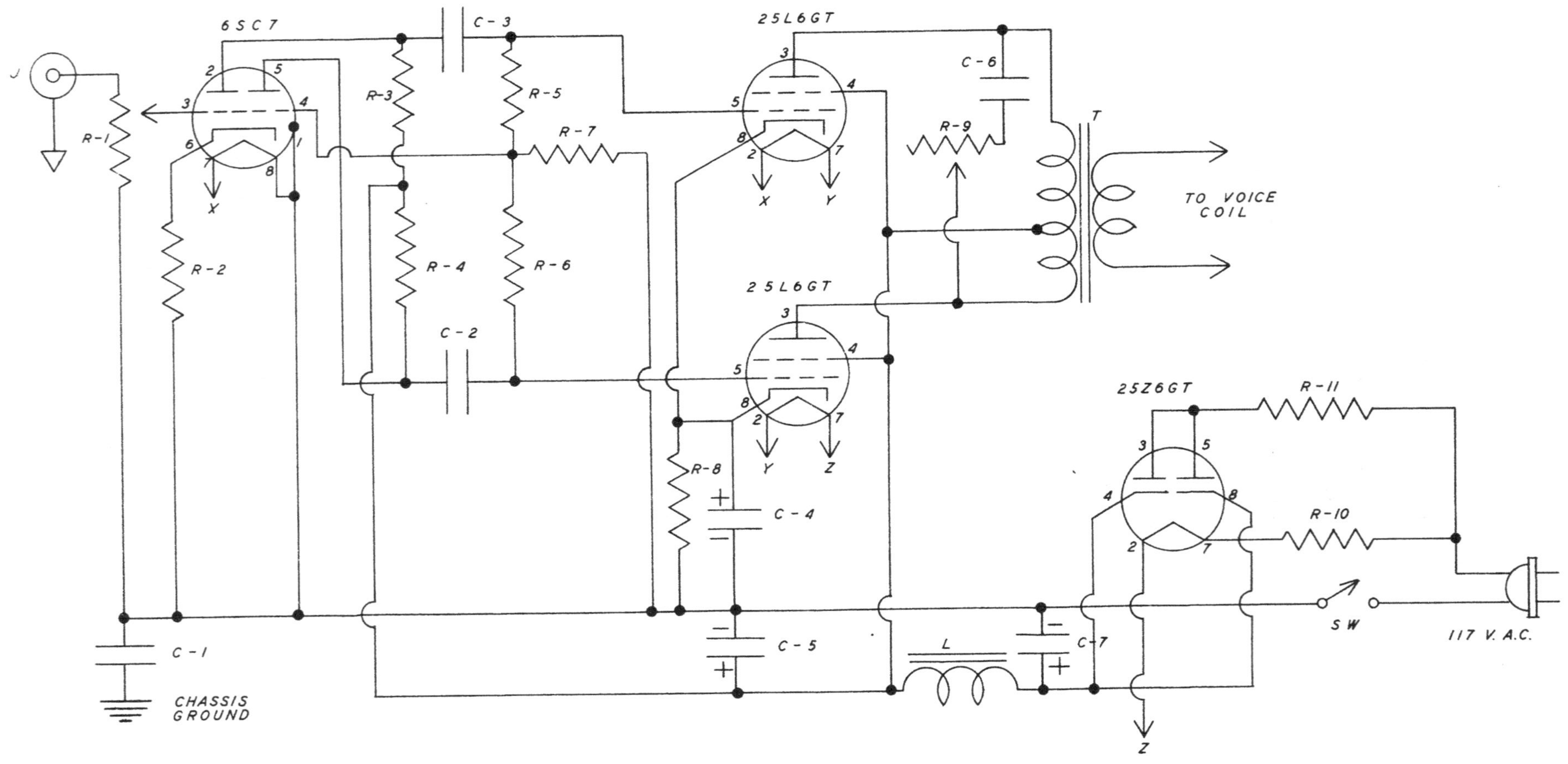

NOTE - ▽ INDICATES COMMON CONNECTIONS ISOLATED FROM CHASSIS

4 TUBE 5 WATT AC/DC AMPLIFIER

The amplifier shown is of the push-pull type, suitable for operation from a phonograph pickup or radio tuner. It is capable of good room volume and fine tone quality due largely to its power output circuit. Having a "transformerless" AC/DC rectifier, the amplifier can be built very compactly, making it ideal for portable purposes. It also operates from either 110 volts A.C. or D.C.

Parts List

C-1	.02 mfd 400 v. paper cond.
C-2	.01 mfd 400 v. paper cond.
C-3	.01 mfd 400 v. paper cond.
C-4	10 mfd 25 v. elec. cond.
C-5	20 mfd 150 v. elec. cond.
C-6	.1 mfd 600 v. paper cond.
C-7	20 mfd 150 v. elec. cond.
R-1	500,000 ohm vol. control
R-2	1500 ohm 1 watt res.
R-3	250,000 ohm ½ watt res.
R-4	250,000 ohm ½ watt res.
R-5	300,000 ohm ½ watt res.
R-6	300,000 ohm ½ watt res.
R-7	150,000 ohm ½ watt res.
R-8	100 ohm 2 watt res.
R-9	100,000 ohm tone control
R-10	125 ohm 25 watt res.
R-11	50 ohm 1 watt res.
T	Output trans: 3000 ohm to to voice coil
L	Filter choke: 100 ohm 125ma
SW	SPST toggle switch
J	Input jack
Sockets:	4 octals

4 TUBE 5 WATT A.C./BAT. P.A. AMPLIFIER

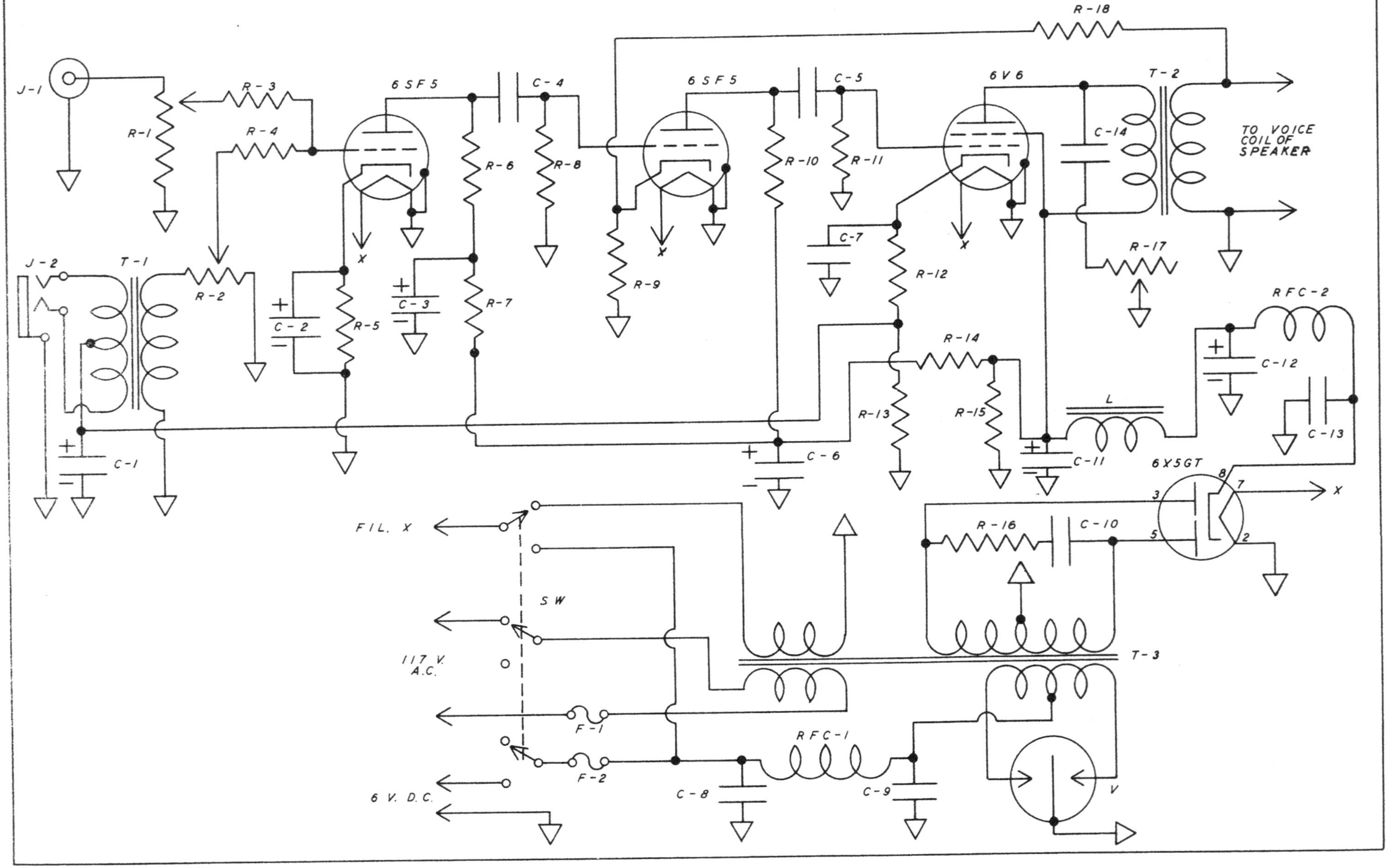

4 TUBE 5 WATT AC/BAT P.A. AMPLIFIER

Here is an amplifier suitable for low power output from a six volt storage battery or 115 volt AC supply. Switching from the 6 volt DC to the 115 volt AC is done by means of sw-1. Both a phonograph pickup and double button carbon microphone can be used simultaneously in the input circuits. Microphone current is obtained from the cathode circuit of the 6V6 tube, and as a result, a sperate excitation battery is not required. Inverse feedback is supplied to the last two amplifier stages for better control.

Building this amplifier requires sufficient chassis area so that the components will not be crowded. Be careful not to place T-1 too close to T-3. All primary leads to T-3 should be as heavy and direct as possible. RFC-1 and 2, 6X5GT, the vibrator, R-16, C-8, C-9, C-10, and C-13, should be located as close to transformer (T-3) as possible.

Parts List

C-1	100 mfd 25 v. elec. cond.
C-2	25 mfd 25 v. elec. cond.
C-3	8 mfd 450 v. elec. cond.
C-4	.05 mfd 600 v. paper cond.
C-5	.05 mfd 600 v. paper cond.
C-6	8 mfd 450 v. elec. cond.
C-7	50 mfd 25 v. elec. cond.
C-8	.5 mfd 200 v. paper cond.
C-9	.5 mfd 200 v. paper cond.
C-10	.01 mfd 1600 v. paper cond.
C-11	16 mfd 450 v. elec. cond.
C-12	16 mfd 450 v. elec. cond.
C-13	.01 mfd 600 v. paper cond.
C-14	.05 mfd 600 v. paper cond.
R-1	500,000 ohm vol. control
R-2	200,000 ohm vol. control
R-3	500,000 ohm ½ watt res.
R-4	2 meg ohm ½ watt res.
R-5	4000 ohm ½ watt res.
R-6	250,000 ohm ½ watt res.
R-7	100,000 ohm ½ watt res.
R-8	500,000 ohm ½ watt res.
R-9	4000 ohm ½ watt res.
R-10	250,000 ohm ½ watt res.
R-11	500,000 ohm ½ watt res.
R-12	200 ohm 2 watt res.
R-13	50 ohm 1 watt res.
R-14	10,000 ohm 1 watt res.
R-15	50,000 ohm 2 watt res.
R-16	100 ohm 1 watt res.
R-17	100,000 ohm tone control
R-18	40,000 ohm ½ watt res.
T-1	Double button mic. trans. 200 ohm c.t. to grid
T-2	Output trans: 5000 ohm to voice coil
T-3	Combination power trans: primary (117 v. A.C. 6-0-6 v.) secondary (300-0-300 v @ 75ma 6.3 v @ 3a)
sw	TPST
RFC-1	10 MH choke
RFC-2	50 turns #12 wire (enamel) on one inch form
V	Non-synchronous vibrator for use with .01 mfd buffer
L	Filter choke: 250 ohm 75ma choke
J-1	Open cir. jack for phono. pickup
J-2	3 cir. jack for carbon mic.
F-1	3 amp fuse
F-2	10 amp fuse
Sockets:	4 octals 1 - 4 prong

4 TUBE 6 WATT A.C. HI-FIDELITY AMPLIFIER

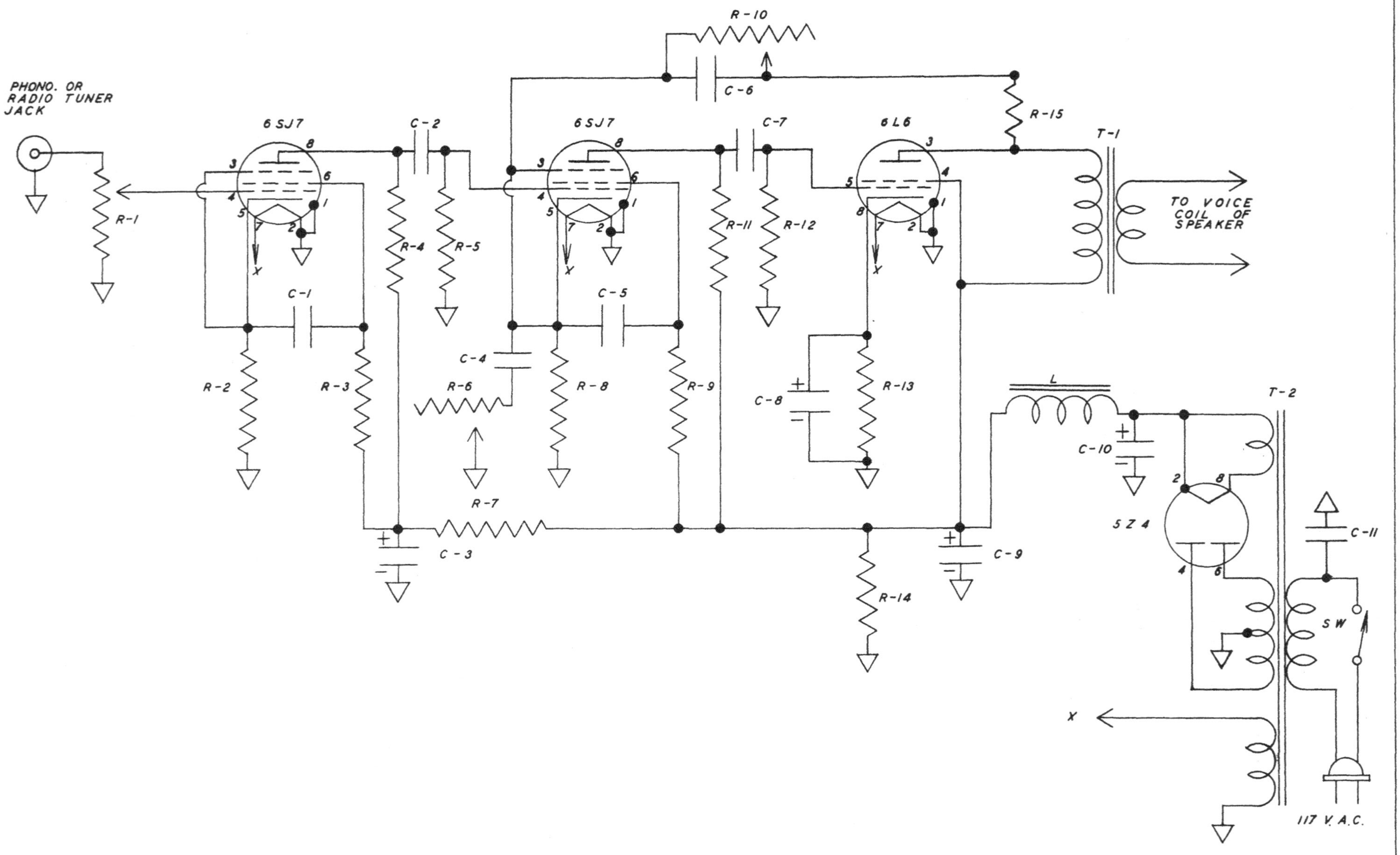

4 TUBE 6 WATT AC HIGH-FIDELITY AMPLIFIER

The above is a high gain, high-fidelity amplifier, capable of excellent volume from a phonograph pickup or radio tuner. It has a unique tone control system arranged in conjunction with the inverse feedback network. This tone control circuit acts to attenuate the amount of feedback by frequency, R-6 and C-4 discriminating on the high frequencies, and R-10 and C-6 on the lows. The balance of the circuit is quite conventional and no particular difficulty should be encountered in its construction.

Parts List

C-1	.25 mfd 600v paper cond.	R-8	1000 ohm $\frac{1}{2}$ watt res.
C-2	.05 mfd 600v paper cond.	R-9	$1\frac{1}{2}$ meg ohm $\frac{1}{2}$ watt res.
C-3	8 mfd 450v elec. cond.	R-10	2 meg ohm bass tone cont.
C-4	.25 mfd 200v paper cond.	R-11	100,000 ohm $\frac{1}{2}$ watt res.
C-5	.25 mfd 600v paper cond.	R-12	250,000 ohm $\frac{1}{2}$ watt res.
C-6	.005 mfd 600v paper cond.	R-13	175 ohm 5 watt res.
C-7	.1 mfd 600v paper cond.	R-14	15,000 ohm 10 watt res.
C-8	50 mfd 25v elec. cond.	R-15	200,000 ohm 1 watt res.
C-9	40 mfd 450v elec. cond.	T-1	Output transformer:
C-10	8 mfd 600v elec. cond.		2500 ohm to voice coil
C-11	.1 mfd 400v paper cond.	T-2	Power transformer:
R-1	500,000 ohm vol. control		300-0-300 v @ 100ma
R-2	1000 ohm $\frac{1}{2}$ watt res.		5 v @ 2a
R-3	$1\frac{1}{2}$ meg ohm $\frac{1}{2}$ watt res.		6.3 v @ 2a
R-4	250,000 ohm $\frac{1}{2}$ watt res.	L	Filter choke: 100 ohm
R-5	1 meg ohm $\frac{1}{2}$ watt res.		125ma
R-6	5000 ohm treble tone cont.	SW	SPST switch on vol. cont.
R-7	50,000 ohm 1 watt res.	Sockets:	4 octals

4 TUBE 6 WATT A.C. HI-FIDELITY AMPLIFIER

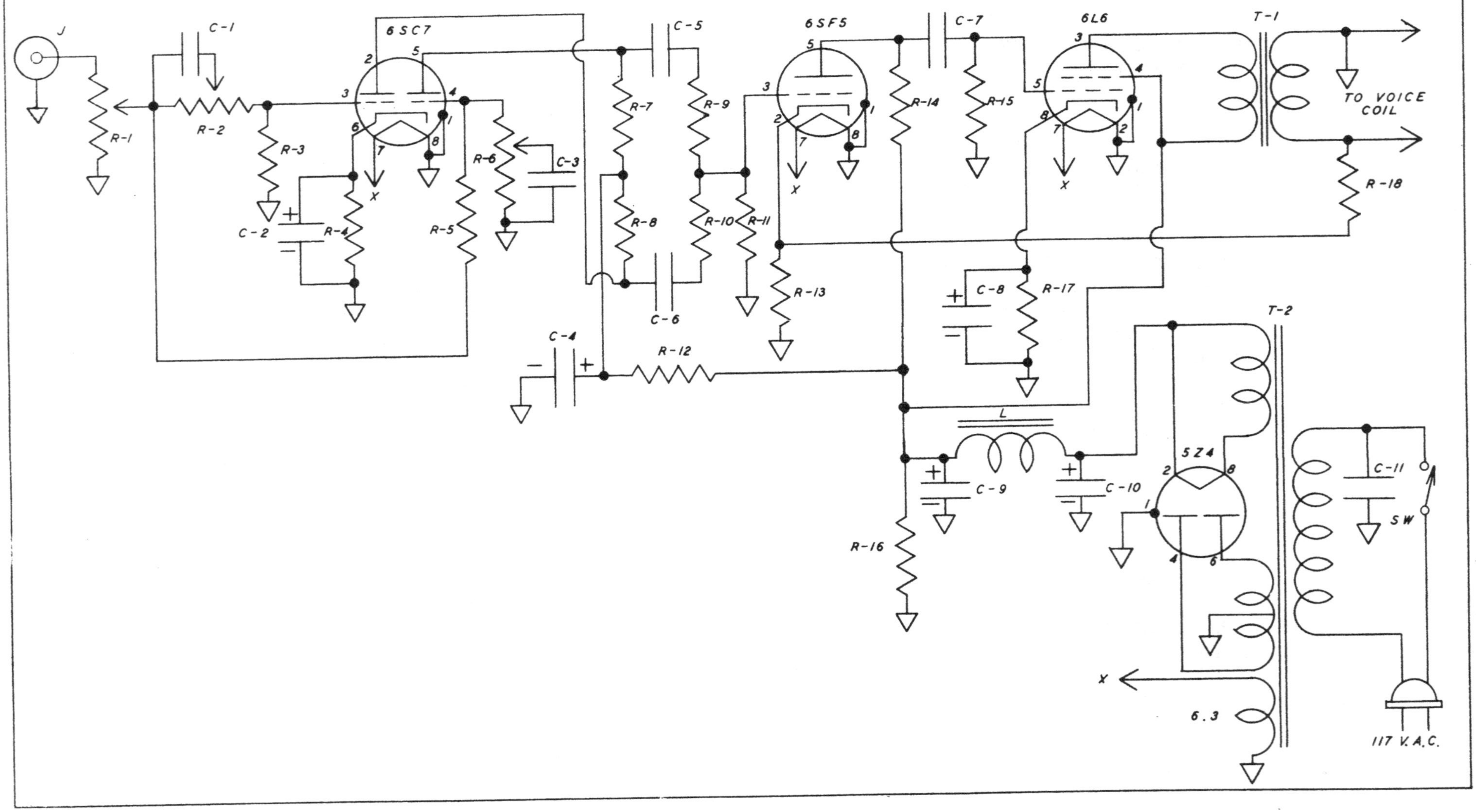

4 TUBE 6 WATT AC HIGH FIDELITY AMPLIFIER

This amplifier is somewhat similar to the previous one, the difference lying in the control circuit. In this instance, R-2 and C-1 control the high frequencies and R-6 and C-3, the low frequencies. The plate circuits of the 6SC7 tube are paralleled, remixing the two tone attenuated outputs. This results in any degree of tone compensation of the two controls being obtained as desired. Phonograph or radio tuner input is recommended for this amplifier.

Parts List

C-1	.0005 mfd 400 v. mica cond.
C-2	25 mfd 25 v. elec. cond.
C-3	.01 mfd 400 v. paper cond.
C-4	8 mfd 400 v. elec. cond.
C-5	.1 mfd 600 v. paper cond.
C-6	.01 mfd 600 v. paper cond.
C-7	.1 mfd 600 v. paper cond.
C-8	50 mfd 50 v. elec. cond.
C-9	40 mfd 450 v. elec. cond.
C-10	8 mfd 600 v. elec. cond.
C-11	.1 mfd 400 v. paper cond.
R-1	1 meg ohm vol. control
R-2	½ meg ohm treble tone cont.
R-3	250,000 ohm ½ watt res.
R-4	1500 ohm 1 watt res.
R-5	¼ meg ohm ½ watt res.
R-6	1 meg ohm bass tone control
R-7	200,000 ohm ½ watt res.
R-8	200,000 ohm ½ watt res.
R-9	200,000 ohm ½ watt res.
R-10	200,000 ohm ½ watt res.
R-11	1 meg ohm ½ watt res.
R-12	20,000 ohm 1 watt res.
R-13	3500 ohm ½ watt res.
R-14	150,000 ohm ½ watt res.
R-15	250,000 ohm ½ watt res.
R-16	20,000 ohm 10 watt res.
R-17	175 ohm 2 watt res.
R-18	40,000 ohm 1 watt res.
T-1	Output trans: 2500 ohm to voice coil 10 watts
T-2	Power trans: 300-0-300 v @ 100ma; 5 v @ 2a; 6.3 v @ 2a
L	Filter choke: 100ma 200 ohm
J	Open circuit input jack
SW	SPST switch on R-1
Sockets:	4 octals

4 TUBE 6 WATT HOME RECORDING AMPLIFIER

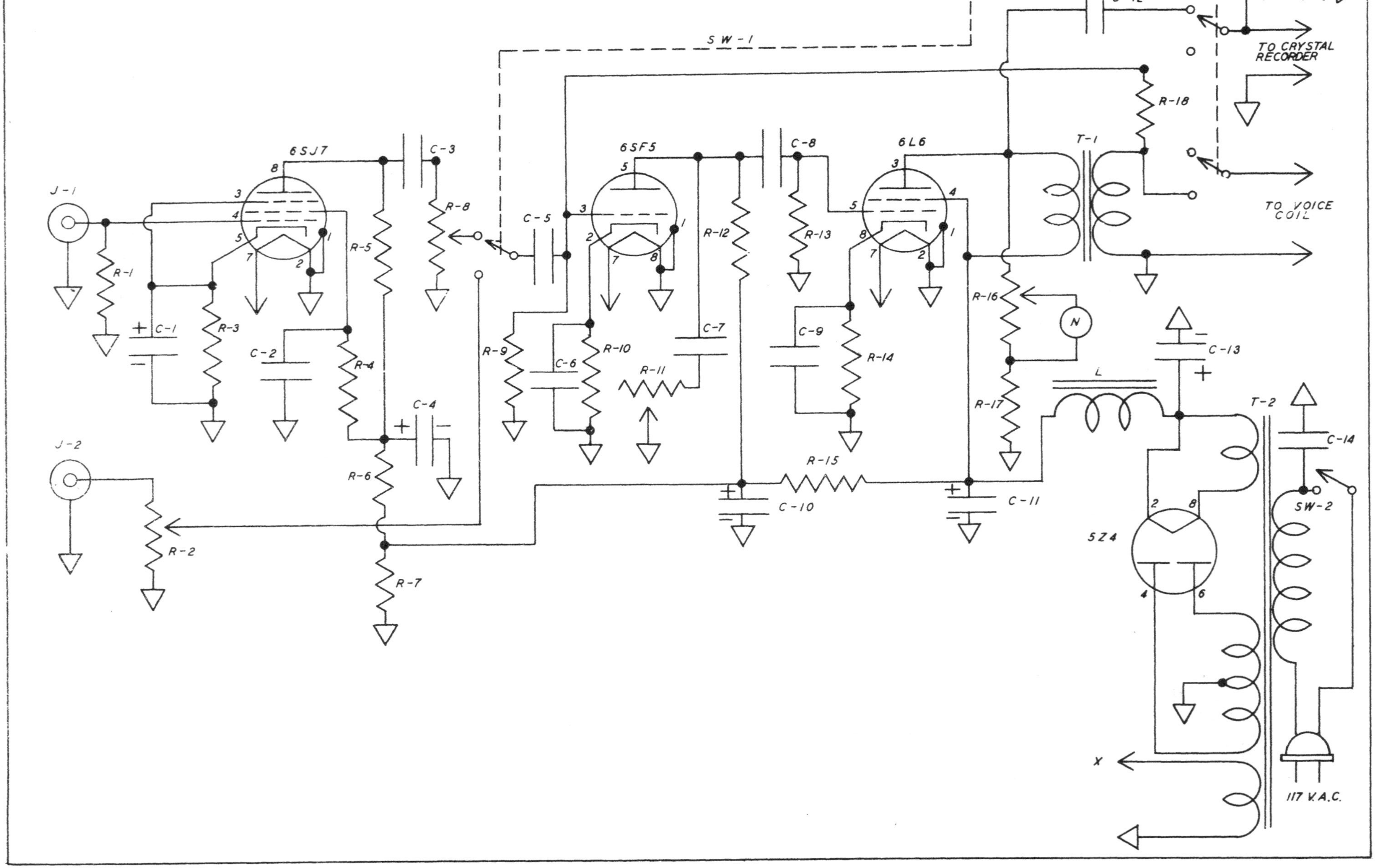

4 TUBE 6 WATT AC HOME RECORDING AMPLIFIER

For an amplifier suitable for home recording and playback, the above amplifier is highly recommended. The input circuits handle a crystal or dynamic high impedance microphone and a phonograph pickup. The output feeds into a speaker or crystal recording head. Switch sw-1 controls the amplifier for either the microphone and recorder, or the phonograph pickup and speaker as desired. When recording, the overmodulation indicator, N, will be very helpful in showing by intensity, the amount of volume necessary to record properly. In initially adjusting this indicator, make test cuts on a recording blank, adjusting the volume control R-7 and indicator R-16 until the recording head just begins to overcut a groove. R-16 is then set for proper operation and need not again be adjusted. A radio tuner may be fed into J-1 for recording radio programs, provided the input signal level does not overload the 6SJ7 tube.

Parts List

C-1	10 mfd 25 v. elec. cond.
C-2	.05 mfd 600 v. paper cond.
C-3	.02 mfd 600 v. paper cond.
C-4	8 mfd 450 v. elec. cond.
C-5	.05 mfd 400 v. paper cond.
C-6	10 mfd 25 v. elec. cond.
C-7	.05 mfd 600 v. paper cond.
C-8	.02 mfd 600 v. paper cond.
C-9	25 mfd 50 v. elec. cond.
C-10	8 mfd 450 v. elec. cond.
C-11	16 mfd 450 v. elec. cond.
C-12	.05 mfd 600 v. paper cond.
C-13	16 mfd 600 v. elec. cond.
C-14	.1 mfd 400 v. paper cond.
R-1	2 meg ohm ½ watt res.
R-2	1 meg ohm vol. control
R-3	3000 ohm ½ watt res.
R-4	100,000 ohm ½ watt res.
R-5	250,000 ohm ½ watt res.
R-6	50,000 ohm 1 watt res.
R-7	25,000 ohm 10 watt res.
R-8	1 meg ohm vol. control
sw-1	3 pole 2 pos. rotary switch
R-9	1 meg ohm ½ watt res.
R-10	5000 ohm ½ watt res.
R-11	50,000 ohm tone control
R-12	250,000 ohm ½ watt res.
R-13	500,000 ohm ½ watt res.
R-14	175 ohm 5 watt res.
R-15	2500 ohm 2 watt res.
R-16	1 meg ohm pot. (linear)
R-17	1 meg ohm ½ watt res.
R-18	2 meg ohm ½ watt res.
R-19	1 meg ohm ½ watt res.
N	¼ watt neon lamp
J-1	Microphone jack
J-2	Phonograph jack
L	Filter choke: 200 ohm-100ma
T-1	Output trans: 2500 ohm to voice coil
T-2	Power trans: 300-0-300 v @ 100ma; 5 v @ 2a; 6.3 v @ 2a
sw-2	SPST toggle switch

4 TUBE 10 WATT A.C. HI-FIDELITY AMPLIFIER

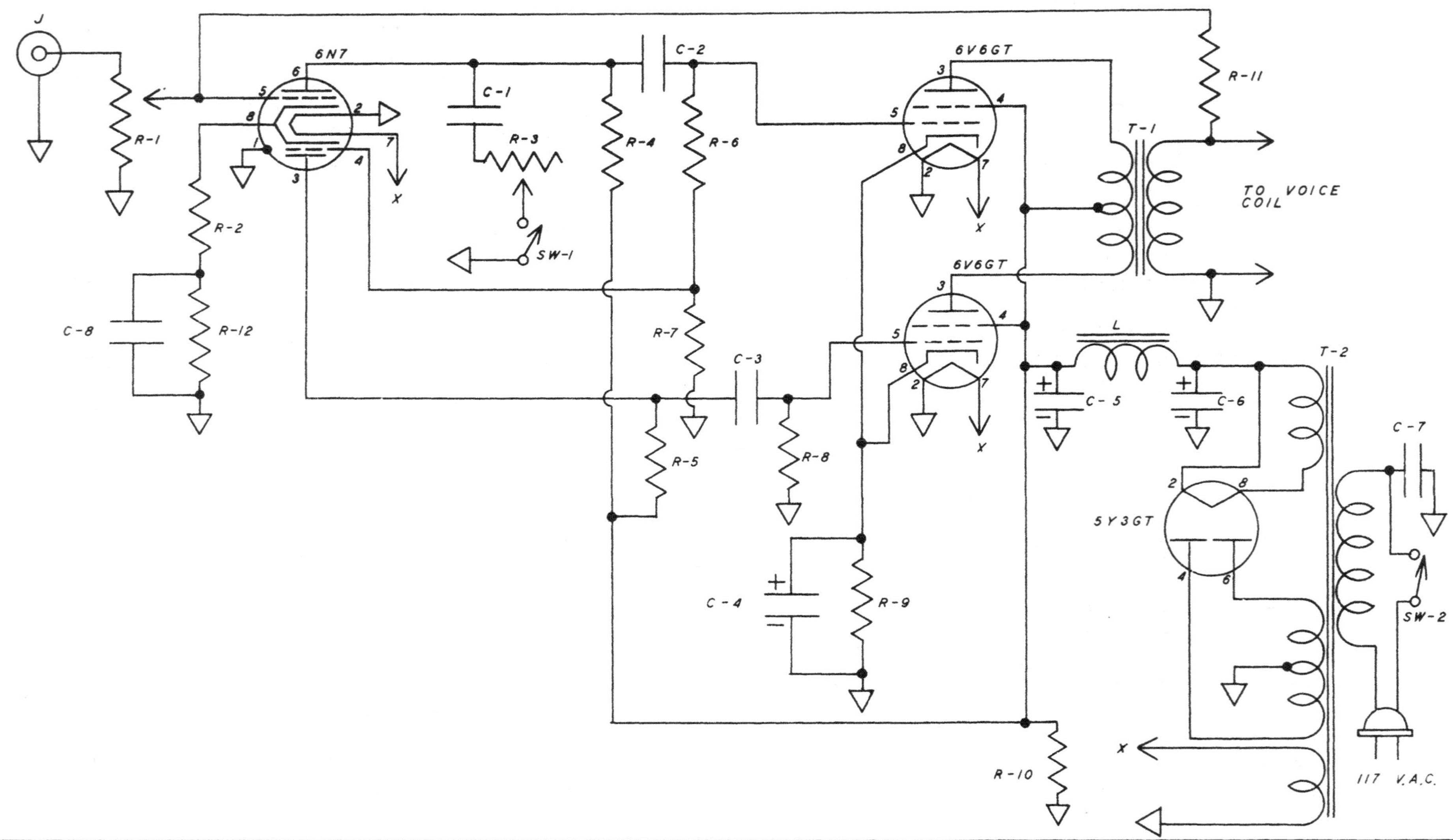

4 TUBE 10 WATT AC HIGH-FIDELITY AMPLIFIER

This is a medium gain phonograph or radio tuner amplifier for high-fidelity reproduction. It features a 6N7 phase inverter and 6V6GT push-pull power output tubes, together with an overall inverse feedback network. A high frequency tone control is provided to reduce the high response when desired. This control is completely cut out by having sw-1 mounted on the control R-3. Of course, high quality parts are necessary to obtain high fidelity response in any amplifier, so care should be taken in their choice. Contruction of this amplifier should offer no particular problem if normal design is followed.

Parts List

C-1	.05 mfd 600 v. paper cond.
C-2	.05 mfd 600 v. paper cond.
C-3	.05 mfd 600 v. paper cond.
C-4	25 mfd 50 v. elec. cond.
C-5	16 mfd 450 v. elec. cond.
C-6	25 mfd 450 v. elec. cond.
C-7	.05 mfd 400 v. paper cond.
C-8	10 mfd 25 v. elec. cond.
R-1	500,000 ohm vol. control
R-2	1500 ohm ½ watt res.
R-3	100,000 ohm tone control
R-4	250,000 ohm ½ watt res.
R-5	250,000 ohm ½ watt res.
R-6	500,000 ohm ½ watt res.
R-7	30,000 ohm ½ watt res.
R-8	500,000 ohm ½ watt res.
R-9	200 ohm 5 watt res.
R-10	25,000 ohm 5 watt res.
R-11	1.5 meg ohm ½ watt res.
R-12	1500 ohm ½ watt res.
T1	Output trans: 8000 ohm to voice coil
T2	Power trans: 300-0-300 v @ 100ma, 5 v @ 2a, 6.3 v @ 2a
L	Filter choke: 300 ohm-100ma - 15 henries
sw-1	SPST switch on tone cont.
sw-2	SPST switch
J	Input jack for radio or phonograph
Sockets:	4 octals

4 TUBE 15 WATT A.C. LOW GAIN AMPLIFIER

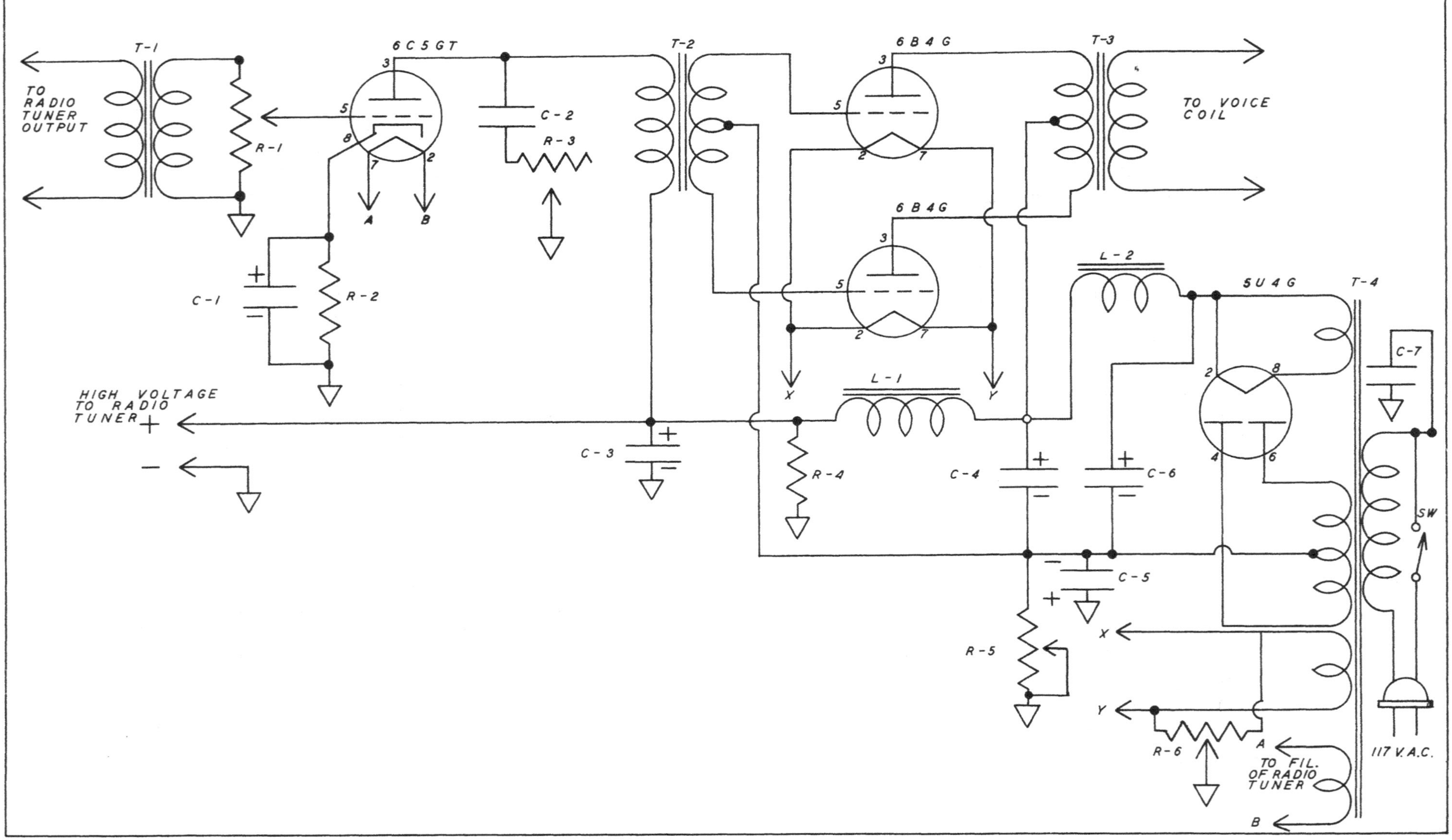

4 TUBE 15 WATT AC LOW GAIN AMPLIFIER

A fully transformer-coupled amplifier is shown, having fixed bias 6B4G push-pull power tubes for distortionless amplification and good power output. It operates from a radio tuner, the power for which is supplied by the amplifier power supply. The fidelity of the amplifier is dependent almost entirely by the quality of the transformers T-1, T-2, and T-3.

It may be difficult to construct this amplifier with its power supply on one chassis, since inductive hum pickup is quite possible. A separate chassis for the power supply will correct this condition. Small amounts of hum can be balanced out of the amplifier by adjusting R-6. In practice, R-5 is adjusted so that a total of 80ma is flowing in the plate circuits of the two 6B4G tubes.

Parts List

C-1	10 mfd 25 v. elec. cond.
C-2	.05 mfd 600 v. paper cond.
C-3	8 mfd 450 v. elec. cond.
C-4	16 mfd 450 v. elec. cond.
C-5	20 mfd 150 v. elec. cond.
C-6	16 mfd 600 v. elec. cond.
C-7	.1 mfd 400 v. paper cond.
R-1	500,000 ohm vol. control
R-2	2500 ohm 1 watt res.
R-3	100,000 ohm tone control
R-4	10,000 ohm 20 watt res.
R-5	800 ohm 20 watt adj. res.
R-6	50 ohm w. w. pot.

T-1	A. F. trans: 3:1 ratio
T-2	P-P A.F. trans: 3:1 ratio
T-3	Output trans: 3000 ohm to voice coil 15 watt
T-4	Power transformer: 400-0-400 v @ 150ma 5 v @ 3a 6.3 v @ 3a 6.3 v @ 3a
L-1	Filter choke: 200 ohm 50ma
L-2	Filter choke: 150 ohm 150ma
SW	SPST toggle switch
Sockets:	4 octals

4 TUBE 3 CHANNEL A.C. PREAMPLIFIER

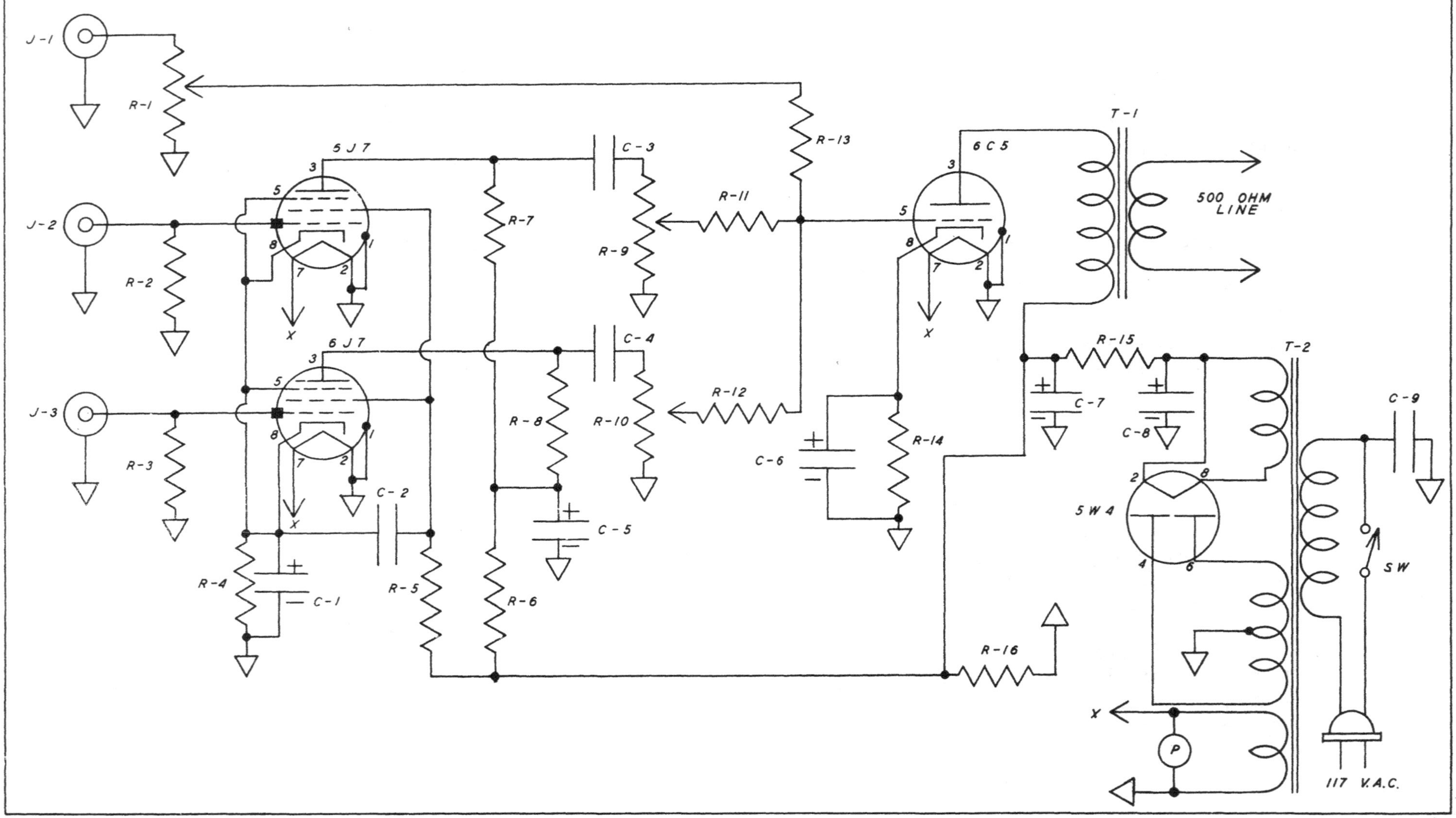

4 TUBE 3 CHANNEL PREAMPLIFIER

A normally low gain amplifier can be operated from a preamplifier similar to the one shown. This designed for a phonograph pickup and two high impedance microphone inputs. These three inputs can be mixed at will to feed the output circuit to the 500 ohm line. Of course, the amplifier following this preamplifier must have an input transformer designed for 500 ohms input.

Good filtering is supplied througnout the high voltage feed circuits, and an entirely seperate power supply makes the preamplifier self sufficient. Little trouble should be encountered in constructing this unit if care is exercised in placing the various parts.

Parts List

C-1	10 mfd 25 v. elec. cond.
C-2	.05 mfd 600 v. paper cond.
C-3	.05 mfd 600 v. paper cond.
C-4	.05 mfd 600 v. paper cond.
C-5	8 mfd 450 v. elec. cond.
C-6	10 mfd 25 v. elec. cond.
C-7	40 mfd 450 v. elec. cond.
C-8	16 mfd 450 v. elec. cond.
C-9	.05 mfd 400 v. paper cond.
R-1	500,000 ohm vol. control
R-2	2 meg ohm ½ watt res.
R-3	2 meg ohm ½ watt res.
R-4	1000 ohm 1 watt res.
R-5	250,000 ohm ½ watt res.
R-6	50,000 ohm 1 watt res.
R-7	250,000 ohm ½ watt res.
R-8	250,000 ohm ½ watt res.
R-9	500,000 ohm vol. control
R-10	500,000 ohm vol. control
R-11	500,000 ohm ½ watt res.
R-12	500,000 ohm ½ watt res.
R-13	250,000 ohm ½ watt res.
R-14	2500 ohm 1 watt res.
R-15	10,000 ohm 10 watt res.
R-16	25,000 ohm 5 watt res.
T-1	Output trans: 10,000 ohm to 500 ohm line
T-2	Power transformer: 250-0-250 v @ 40ma; 5 v @ 2a; 6.3 v @ 1a
J-1	Phonograph jack
J-2	Microphone jack
J-3	Microphone jack
SW	SPST toggle switch
P	6.3 v. panel light
Sockets:	4 octals

5 TUBE 10 CLASS-B 6 VOLT D.C. AMPLIFIER

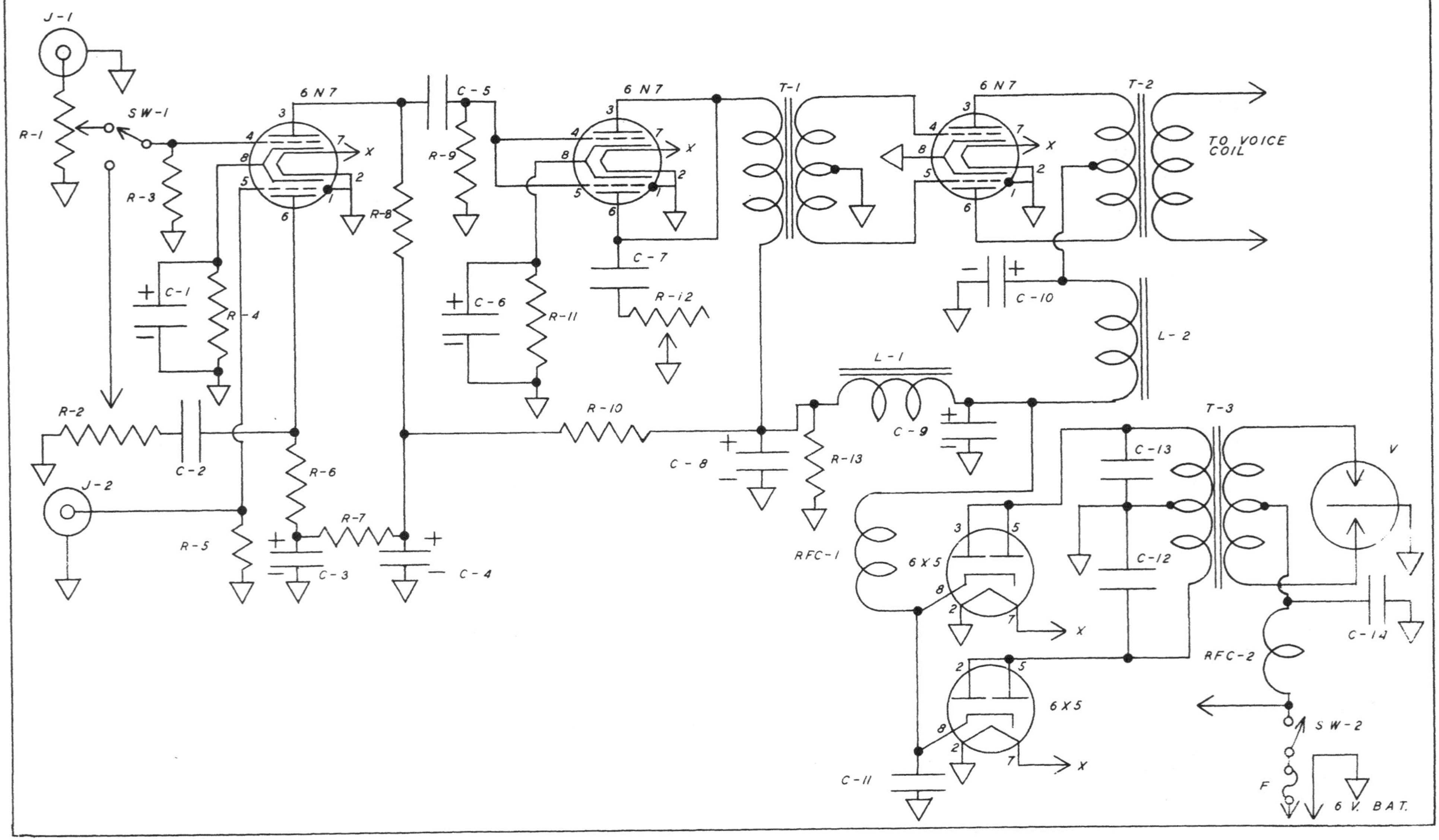

5 TUBE 10 WATT CLASS-B 6 VOLT DC AMPLIFIER

For a six volt battery operated amplifier, this arrangement will provide good volume and power output from a crystal or dynamic microphone and phonograph pickup. It is satisfactory for use in a sound truck or car when too great a speaker coverage is not demanded. Three 6N7 and two 6X5 tubes are used throughout the amplifier, simplifying the number of replacement tubes to carry. Be certain a heavy duty vibrator is used, as the primary current consumption may be 10 amperes. Since it is seldom necessary to operate the phonograph and microphone simultaneously, a switch sw-1, selects the current desired.

If, after completing the amplifier you are dissatisfied with the tone, try shunting the secondary winding of T-1 to ground with equal resistors of 100,000 ohms, ½ watt each. In any class "B" amplifier, the tone is always improved when the power tubes are driven fairly hard, consequently the amplifier sounds better when the volume is high. Objectionable vibrator "hash" is generally due to insufficient shielding or poor grounds, and at times, can only be corrected by trial and error.

Parts List

C-1	25 mfd 25v elec. cond.
C-2	.02 mfd 600v paper cond.
C-3	8 mfd 450v elec. cond.
C-4	8 mfd 450v elec. cond.
C-5	.05 mfd 600v paper cond.
C-6	25 mfd 25v elec. cond.
C-7	.02 mfd 600v paper cond.
C-8	16 mfd 450v elec. cond.
C-9	8 mfd 450v elec. cond.
C-10	40 mfd 450v elec. cond.
C-11	.01 mfd 600v paper cond.
C-12	.01 mfd 1500v paper cond.
C-13	.01 mfd 1500v paper cond.
C-14	.5 mfd 400v paper cond.
R-1	500,000 ohm vol. control
R-2	500,000 ohm vol. control
R-3	1 meg ohm ½ watt res.
R-4	3000 ohm 1 watt res.
R-5	2 meg ohm ½ watt res.
R-6	250,000 ohm ½ watt res.
R-7	100,000 ohm ½ watt res.
R-8	250,000 ohm ½ watt res.
R-9	500,000 ohm ½ watt res.
R-10	10,000 ohm 1 watt res.
R-11	1500 ohm 2 watt res.
R-12	100,000 ohm tone cont.
R-13	20,000 ohm 10 watt res.
F	20 amp fuse
T-1	Class B input trans: 6N7 parallel to 6N7 P.P.
T-2	Class B output trans: 10w 8000 ohm to v.c.
T-3	Vibrator power trans: 300-0-300 v @ 100ma 6-0-6 v primary
V	4 prong H.D. non-sync. vibrator
RFC-1	10MH R.F. choke
RFC-2	50 turns #12 enam. wire on ½" form
L-1	Filter choke: 75ma 250 ohm
L-2	Filter choke: 75ma 250 ohm
sw-1	SPST phono. micro. toggle switch
sw-2	SPST power switch 20a
J-1	Open cir. phono. jack
J-2	Open cir. micro. jack
Sockets:	5 octals, 1-4 prong

5 TUBE 13 WATT A.C. HI-FIDELITY AMPLIFIER

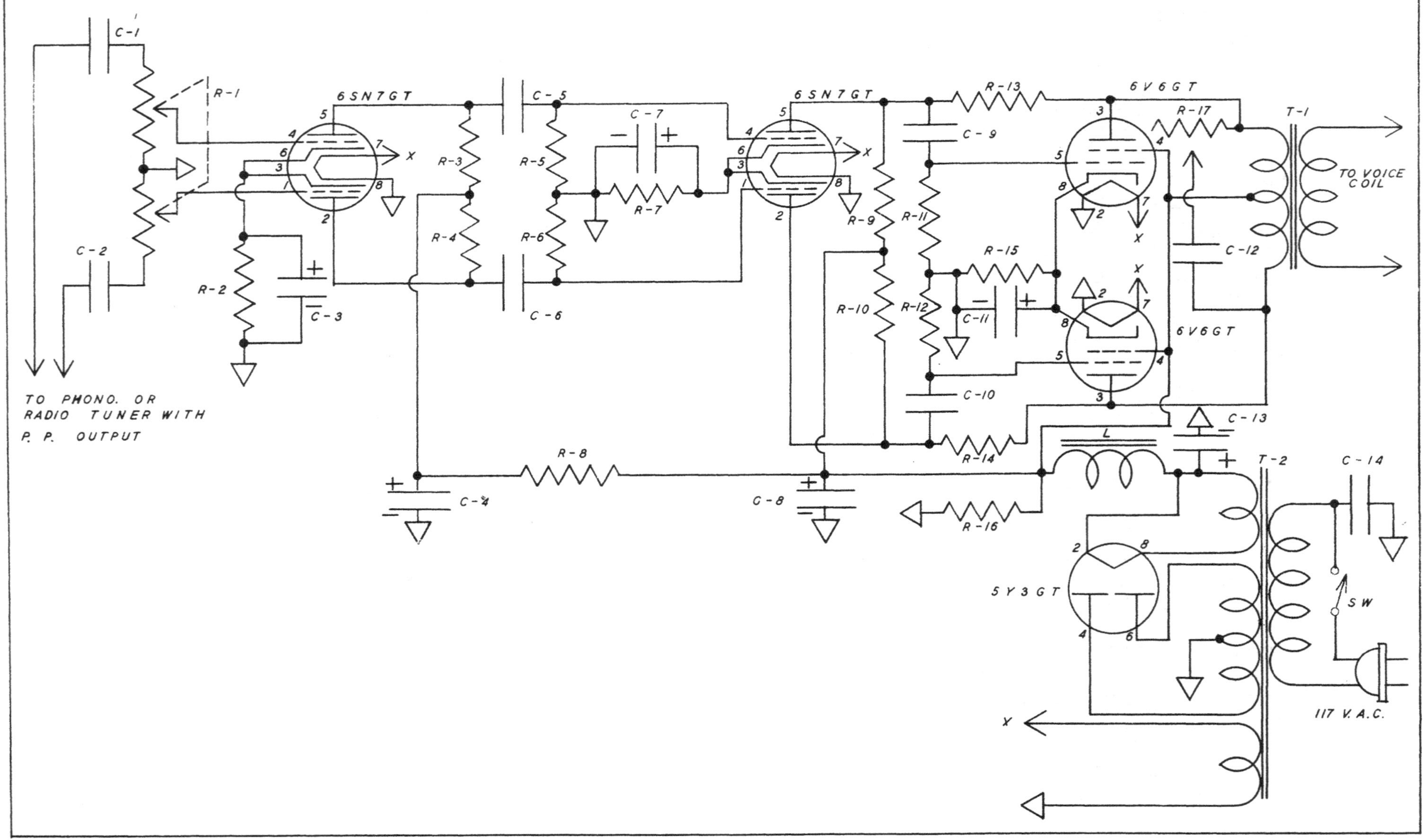

5 TUBE 13 WATT AC HIGH-FIDELITY AMPLIFIER

Here is an A.C. amplifier of very fine tone quality, having all push-pull stages. Likewise, it is suitable for operation from either a phonograph pickup or radio tuner. When connected to a crystal pickup unit, the input leads connect directly to the pickup. If a radio tuner is used, a high quality 1:1 ratio input transformer should be used, having a push-pull secondary in order to get proper push-pull input. Inverse feedback is provided through resistors R-14 and R-13.

Generally, cathode by-pass condensers are not necessary in a push-pull amplifier, but are here specified to compensate for slight differences of emission between the various pairs of push-pull tubes. The volume control, R-1, is a dual affair, giving equal variations of both grid resistances for each degree of rotation. The tone control, R-17, effectively short-circuits the higher audio frequencies as the resistance is reduced. When connected to a good speaker system, this amplifier should give excellent fidelity and volume.

Parts List

C-1	.05 mfd 600 v. paper cond.
C-2	.05 mfd 600 v. paper cond.
C-3	25 mfd 25 v. elec. cond.
C-4	8 mfd 450 v. elec. cond.
C-5	.05 mfd 600 v. paper cond.
C-6	.05 mfd 600 v. paper cond.
C-7	25 mfd 50 v. elec. cond.
C-8	40 mfd 450 v. elec. cond.
C-9	.05 mfd 600 v. paper cond.
C-10	.05 mfd 600 v. paper cond.
C-11	25 mfd 50 v. elec. cond.
C-12	.02 mfd 600 v. paper cond.
C-13	16 mfd 600 v. elec. cond.
C-14	.1 mfd 400 v. paper cond.
R-1	Dual 500,000 ohm vol. cont.
R-2	2500 ohm 1 watt res.
R-3	75,000 ohm ½ watt res.
R-4	75,000 ohm ½ watt res.
R-5	500,000 ohm ½ watt res.
R-6	500,000 ohm ½ watt res.
R-7	3000 ohm 1 watt res.
R-8	10,000 ohm 1 watt res.
R-9	75,000 ohm ½ watt res.
R-10	75,000 ohm ½ watt res.
R-11	500,000 ohm ½ watt res.
R-12	500,000 ohm ½ watt res.
R-13	1 meg ohm 1 watt res.
R-14	1 meg ohm 1 watt res.
R-15	200 ohm 5 watt res.
R-16	20,000 ohm 10 watt res.
R-17	100,000 ohm tone control
L	Filter choke: 200 ohm 100ma
T-1	Output trans: 8000 ohm to voice coil 15 watt
T-2	Power trans: 325-0-325 v @ 100ma; 5 v @ 2a; 6.3 v @ 2a
SW	SPST switch
Sockets:	5 octals

5 TUBE 13 WATT A.C. HI-FIDELITY AMPLIFIER

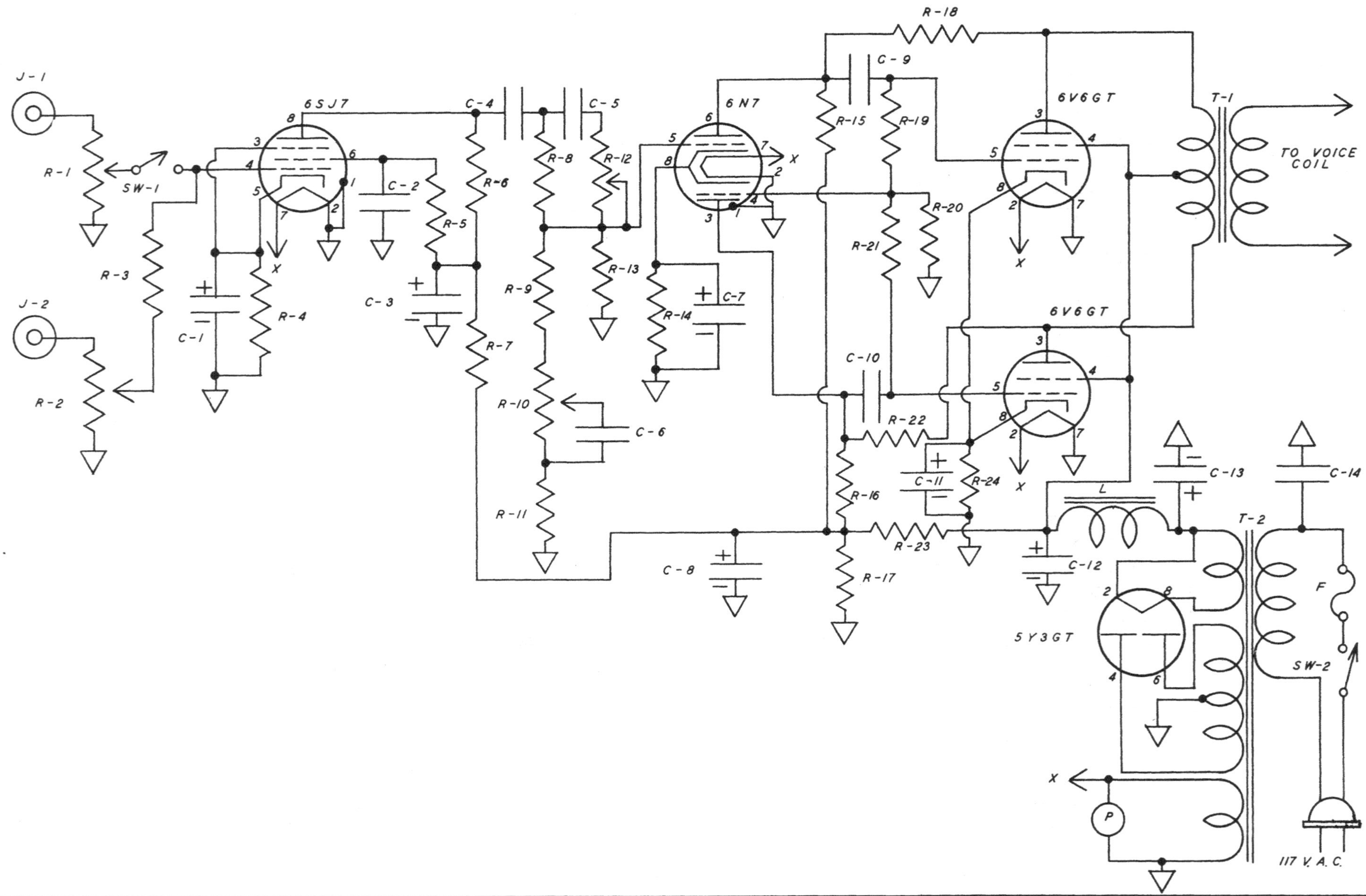

5 TUBE 13 WATT AC HIGH-FIDELITY P.A. AMPLIFIER

This public address amplifier, having a phonograph and high impedance microphone inputs, is capable of very good fidelity and output. It features separate bass (R-10) and treble (R-12) tone controls, and a separate microphone (R-1) and phonograph (R-2) volume controls. When it is desired to operate the phonograph without the microphone, R-1 should be turned to the left until sw-1 opens so that R-1 will not have any shunting effect on the grid of the 6SJ7 tube. However, the position of R-2 has no effect on the microphone volume.

Parts List

C-1	10 mfd 25 v. elec. cond.
C-2	.1 mfd 600 v. paper cond.
C-3	8 mfd 450 v. elec. cond.
C-4	.05 mfd 600 v. paper cond.
C-5	.0005 mfd mica cond.
C-6	.01 mfd 400 v. paper cond.
C-7	10 mfd 25 v. elec. cond.
C-8	8 mfd 450 v. elec. cond.
C-9	.05 mfd 600 v. paper cond.
C-10	.05 mfd 600 v. paper cond.
C-11	25 mfd 25 v. elec. cond.
C-12	40 mfd 450 v. elec. cond.
C-13	8 mfd 600 v. elec. cond.
C-14	.1 mfd 400 v. paper cond.
R-1	1 meg ohm mic. vol. cont.
R-2	½ meg ohm phono. vol. cont.
R-3	1 meg ohm ½ watt res.
R-4	2000 ohm ½ watt res.
R-5	1 meg ohm ½ watt res.
R-6	200,000 ohm ½ watt res.
R-7	50,000 ohm ½ watt res.
R-8	250,000 ohm ½ watt res.
R-9	10,000 ohm ½ watt res.
R-10	5 meg ohm bass tone cont.
R-11	15,000 ohm ½ watt res.
P	Jewel indicator 6 v. lamp
Sockets:	5 octals
R-12	1 meg ohm treble tone cont.
R-13	500,000 ohm ½ watt res.
R-14	2500 ohm 1 watt res.
R-15	100,000 ohm ½ watt res.
R-16	100,000 ohm ½ watt res.
R-17	20,000 ohm 5 watt res.
R-18	1 meg ohm ½ watt res.
R-19	1 meg ohm ½ watt res.
R-20	400,000 ohm ½ watt res.
R-21	1 meg ohm ½ watt res.
R-22	100,000 ohm ½ watt res.
R-23	5000 ohm 5 watt res.
R-24	200 ohm 2 watt res.
T-1	Output trans: 8000 ohm to voice coil
T-2	Power trans: 300-0-300 @ 120ma; 5 v @ 2A; 6.3 v @ 2A
L	Filter choke: 200 ohm 120ma
J-1	Open cir. mic. jack
J-2	Open cir. phono. jack
F	3 amp fuse
sw-1	SPST mic. switch on R1
sw-2	SPST AC switch on R2, or as separate toggle

5 TUBE 15 WATT QUICK HEATING PHONOGRAPH AMPLIFIER

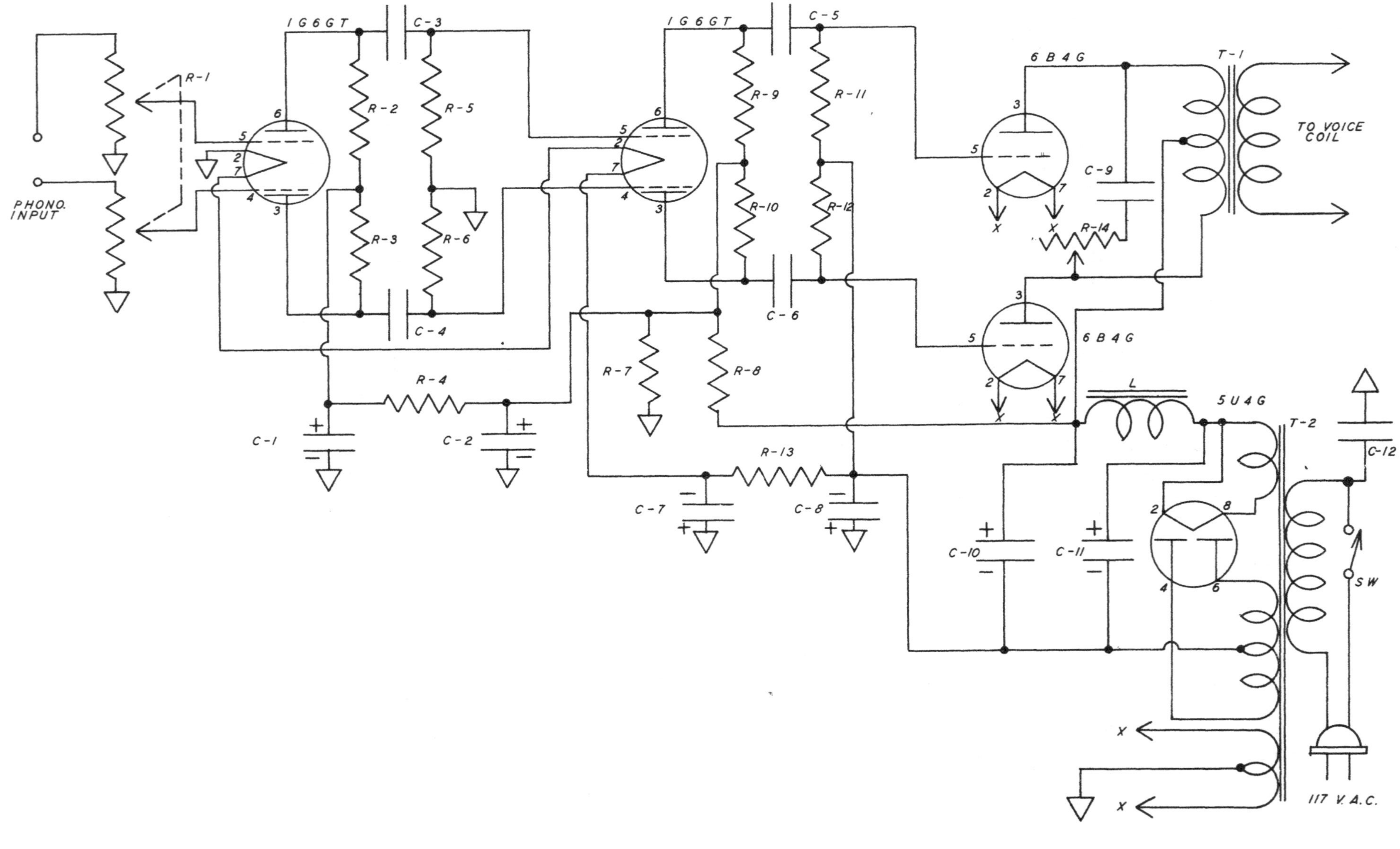

5 TUBE 15 WATT QUICK HEATING PHONOGRAPH AMPLIFIER

The above is a quick heating, all push-pull, phonograph amplifier, capable of good volume and tone. It is especially desireable for use in intermittent automatic phonograph players--commonly called a "juke box." When turned on, the amplifier is fully ready to operate within 15 seconds, as all tubes operate as a fixed bias stage and together with the 300 volts on their plates, give very strong output power without distortion. A dual volume control, R-1 and R-2, adjusts the input level and a tone control (R-14) sets the high frequency response.

Parts List

C-1	8 mfd 450 v. elec. cond.
C-2	8 mfd 450 v. elec. cond.
C-3	.05 mfd 600 v. paper cond.
C-4	.05 mfd 600 v. paper cond.
C-5	.1 mfd 600 v. paper cond.
C-6	.1 mfd 600 v. paper cond.
C-7	500 mfd 6 v. elec. cond.
C-8	40 mfd 150 v. elec. cond.
C-9	.1 mfd 600 v. paper cond.
C-10	25 mfd 600 v. elec. cond.
C-11	8 mfd 600 v. elec. cond.
C-12	.1 mfd 400 v. paper cond.
R-1	Dual 500,000 ohm vol. cont.
R-2	250,000 ohm ½ watt res.
R-3	250,000 ohm ½ watt res.
R-4	100,000 ohm 1 watt res.
R-5	500,000 ohm ½ watt res.
R-6	500,000 ohm ½ watt res.
R-7	10,000 ohm 10 watt res.
R-8	12,500 ohm 10 watt res.
R-9	100,000 ohm ½ watt res.
R-10	100,000 ohm ½ watt res.
R-11	100,000 ohm 1 watt res.
R-12	100,000 ohm 1 watt res.
R-13	550 ohm 20 watt res.
R-14	250,000 ohm tone control
T-1	Output trans: 15 watts 3000 ohm to voice coil
T-2	Power transformer: 325-0-325 v @ 110ma 5 v @ 2a 6.3 v @ 3a
L	Filter choke: 20Hy 110ma
SW	SPST toggle switch
Sockets:	5 octals

5 TUBE 35 WATT TRANSMITTER MODULATOR

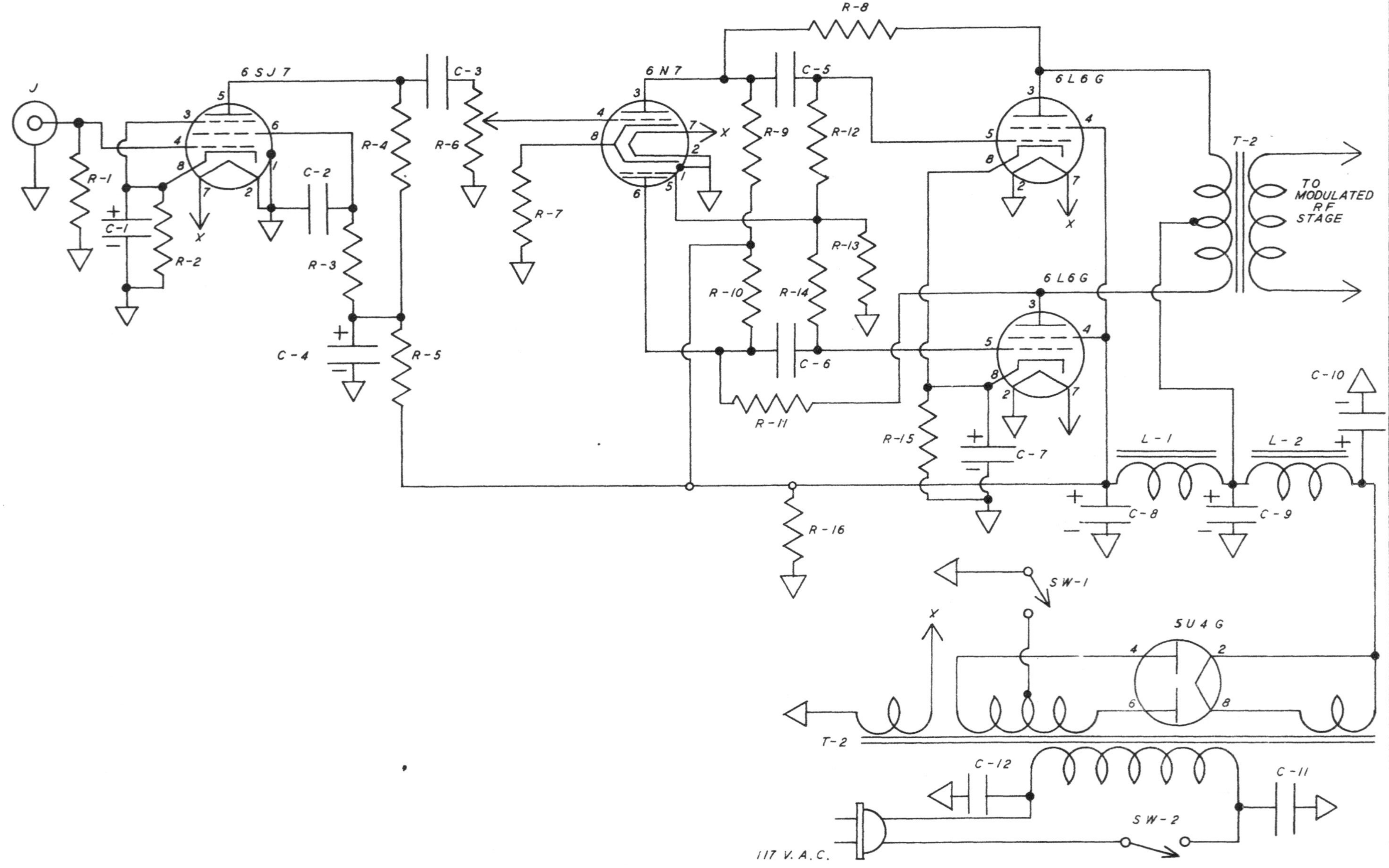

5 TUBE 35 WATT TRANSMITTER MODULATOR

This 35 watt amplifier is quite suitable for modulating an amateur or commercial 50 to 75 watt transmitter. It is designed for a single crystal or dynamic microphone input and has considerable over-all gain. A 6SJ7 high gain stage operates into a 6N7 phase inverter, and thence into the push-pull 6L6G tubes for high power. A load should always be applied to the output transformer secondary to protect it against accidental burnout.

The major problem in building a transmitter modulator is in keeping radio frequency currents out of the input of the amplifier. This is usually accomplished by inserting a RF choke in series with the input to the amplifier and grounding the amplifier chassis to a waterpipe or some other suitable ground.

Parts List

C-1	10 mfd 25 v. elec. cond.
C-2	.05 mfd 600 v. paper cond.
C-3	.01 mfd 600 v. paper cond.
C-4	8 mfd 450 v. elec. cond.
C-5	.01 mfd 600 v. paper cond.
C-6	.01 mfd 600 v. paper cond.
C-7	25 mfd 50 v. elec. cond.
C-8	16 mfd 450 v. elec. cond.
C-9	25 mfd 450 v. elec. cond.
C-10	16 mfd 600 v. elec. cond.
C-11	.02 mfd 400 v. paper cond.
C-12	.02 mfd 400 v. paper cond.
R-1	2 meg ohm ½ watt res.
R-2	3000 ohm ½ watt res.
R-3	1 meg ohm ½ watt res.
R-4	250,000 ohm ½ watt res.
R-5	50,000 ohm 1 watt res.
R-6	500,000 ohm vol. control
R-7	4000 ohm 1 watt res.
R-8	1 meg ohm ½ watt res.
R-9	100,000 ohm ½ watt res.
R-10	100,000 ohm ½ watt res.
R-11	1 meg ohm ½ watt res.
R-12	250,000 ohm ½ watt res.
R-13	100,000 ohm ½ watt res.
R-14	250,000 ohm ½ watt res.
R-15	150 ohm 10 watt res.
R-16	15,000 ohm 25 watt res.
L-1	Filter choke: 400 ohm-50ma
L-2	Filter choke: 100 ohm-200ma
T-1	Output trans: 25 watts 5000 ohm to tapped sec.
T-2	Power transformer: 400-0-400 v @ 150ma; 5 v @ 3a; 6.3 v @ 3½a
J	Open circuit mic. jack
sw-1	SPST communication switch
sw-2	SPST power switch
Sockets:	5 octals

5 TUBE 30 WATT A.C. P.A. AMPLIFIER

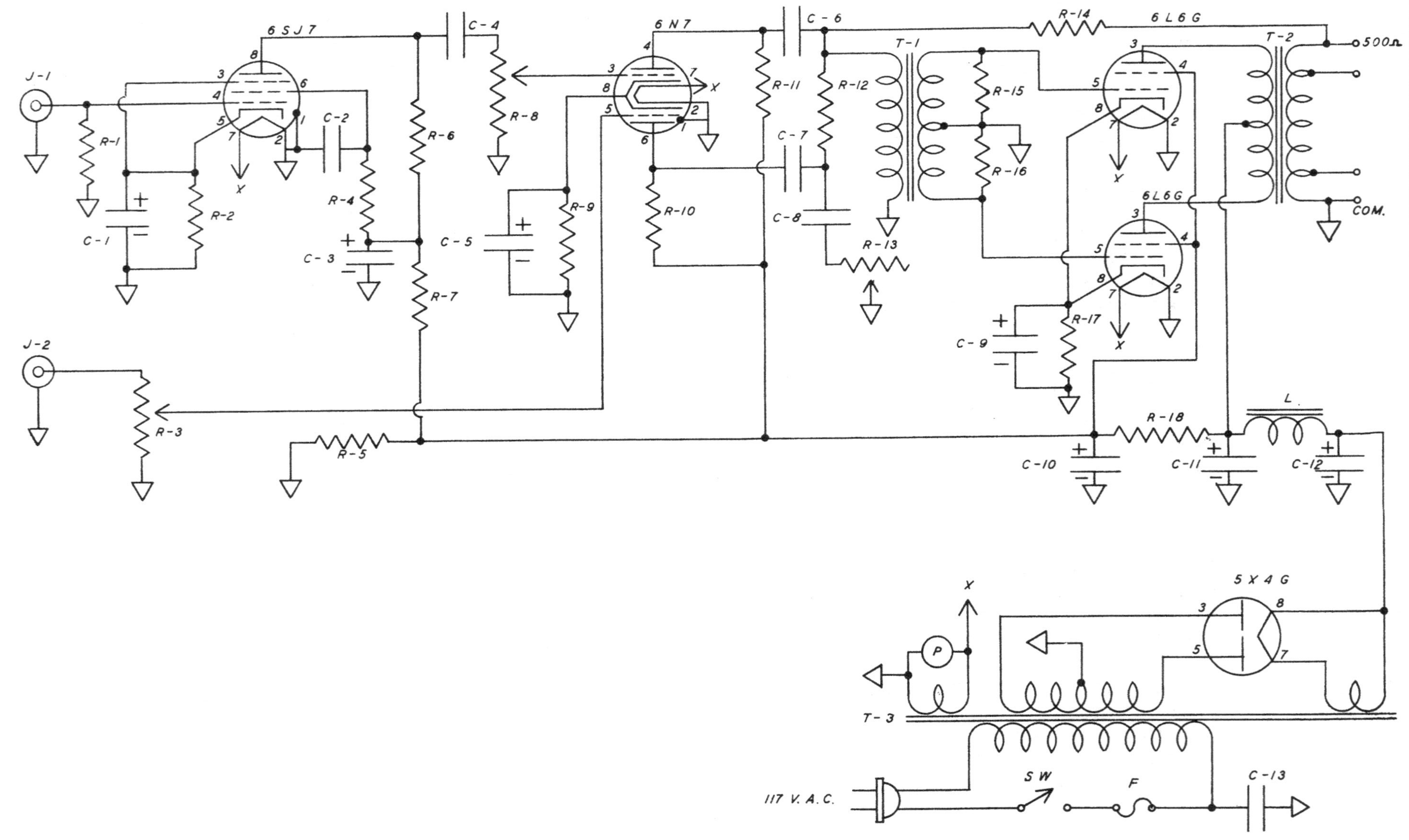

5 TUBE 30 WATT AC P.A. HIGH-FIDELITY AMPLIFIER

This is an amplifier similar to the previous one, but is more suitable for public address work. Instead of a 6N7 phase inverter, the 6N7 tube is used as an electronic mixer feeding into a push-pull input transformer. This type of mixer allows any setting of either volume control (R-3 or R-8) to be made without affecting the other. A microphone and phonograph input can be used simultaneously, and any type of speaker or speakers matched to the output. Inverse feedback aids in the fidelity of this amplifier, giving it smooth operation.

Parts List

C-1	25 mfd 25 v. elec. cond.
C-2	.1 mfd 600 v. paper cond.
C-3	8 mfd 450 v. elec. cond.
C-4	.05 mfd 600 v. paper cond.
C-5	25 mfd 25 v. elec. cond.
C-6	.25 mfd 600 v. paper cond.
C-7	.25 mfd 600 v. paper cond.
C-8	.02 mfd 600 v. paper cond.
C-9	25 mfd 50 v. elec. cond.
C-10	16 mfd 450 v. elec. cond.
C-11	25 mfd 450 v. elec. cond.
C-12	16 mfd 600 v. elec. cond.
C-13	.1 mfd 400 v. paper cond.
R-1	2 meg ohm ½ watt res.
R-2	1000 ohm ½ watt res.
R-3	500,000 ohm vol. cont.
R-4	1 meg ohm ½ watt res.
R-5	20,000 ohm 10 watt res.
R-6	150,000 ohm ½ watt res.
R-7	50,000 ohm 1 watt res.
R-8	500,000 ohm vol. control
R-9	2000 ohm 1 watt res.
R-10	100,000 ohm ½ watt res.
R-11	100,000 ohm ½ watt res.
R-12	100,000 ohm ½ watt res.
R-13	50,000 ohm tone control
R-14	500,000 ohm ½ watt res.
R-15	200,000 ohm ½ watt res.
R-16	200,000 ohm ½ watt res.
R-17	200 ohm 10 watt res.
R-18	3000 ohm 10 watt res.
L	Filter choke: 150 ohm-200ma
T-1	AF trans: 2:1 ratio
T-2	Output trans: 30 watt 6600 ohm to multi-tap sec.
T-3	Power trans: 375-0-375 v @ 200ma, 5 v @ 3a, 6.3 v @ 3a
J-1	Microphone jack
J-2	Phonograph jack
F	3 amp fuse
P	6.3 v pilot light
SW	SPST switch
Sockets:	5 octals

6 TUBE 6 WATT TWO CHANNEL AMPLIFIER

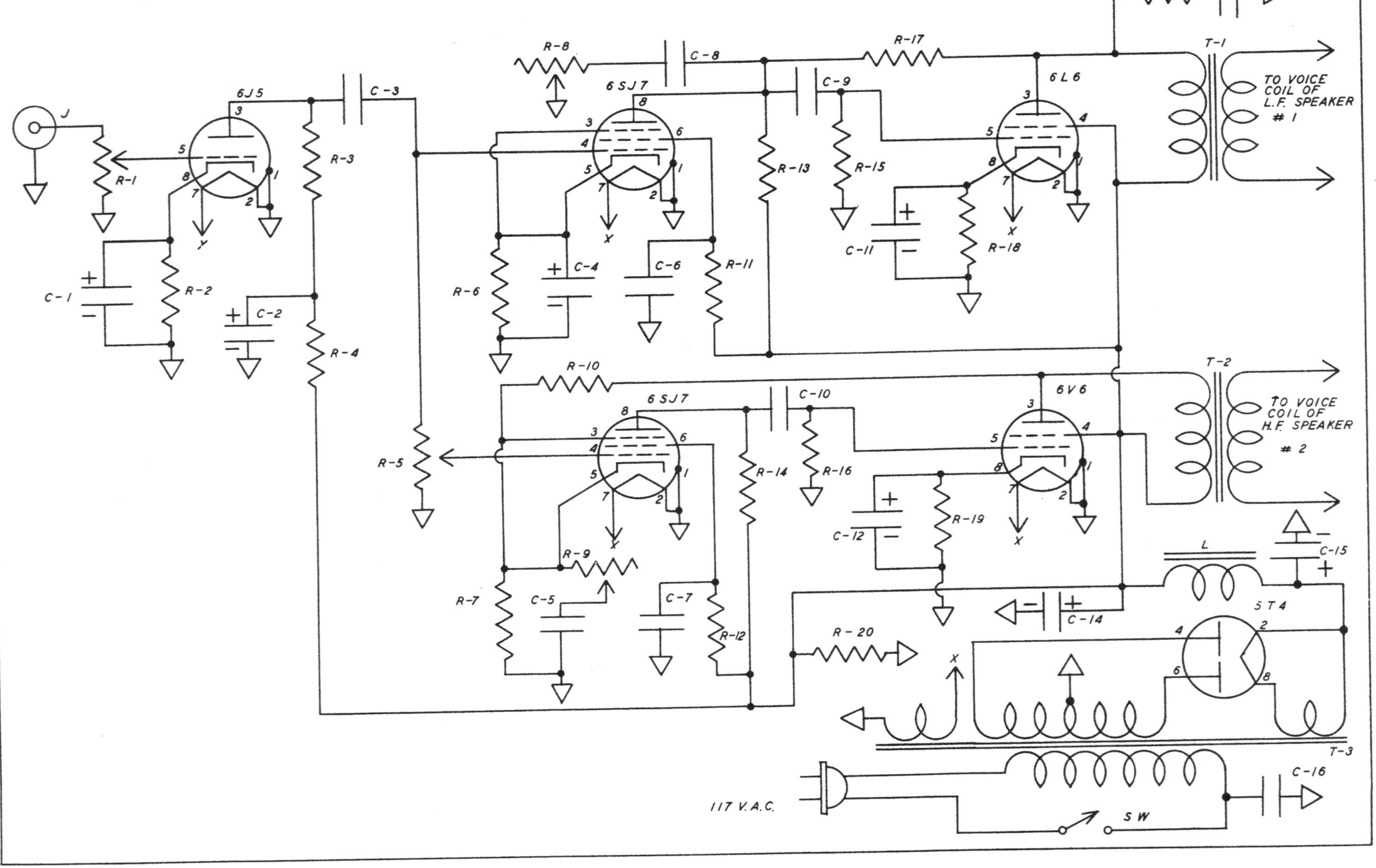

6 TUBE 6 WATT AC 2 CHANNEL AMPLIFIER

Here is a two channel amplifier suitable to operate from a radio tuner or phonograph pickup. One channel is designed to pass the lower audio frequencies while the other passes the higher frequencies. The degree of volume in each channel is controlled by tone controls R-8 and R-9 so that the constructor may compensate for room conditions. R-1 is the overall volume control for the system, while R-5 is preliminarily adjusted for tone balance. With one large speaker for low frequencies and a small one for high frequencies, the system is capable of excellent fidelity and volume.

Parts List

C-1	25 mfd 25 v. elec. cond.
C-2	8 mfd 450 v. elec. cond.
C-3	.05 mfd 600 v. paper cond.
C-4	25 mfd 25 v. elec. cond.
C-5	.5 mfd 400 v. paper cond.
C-6	.25 mfd 600 v. paper cond.
C-7	.1 mfd 600 v. paper cond.
C-8	.01 mfd 600 v. paper cond.
C-9	.1 mfd 600 v. paper cond.
C-10	.005 mfd 600 v. paper cond.
C-11	50 mfd 50 v. elec. cond.
C-12	10 mfd 50 v. elec. cond.
C-13	.01 mfd 600 v. paper cond.
C-14	40 mfd 450 v. elec. cond.
C-15	16 mfd 600 v. elec. cond.
C-16	.1 mfd 400 v. paper cond.
R-1	500,000 ohm vol. control
R-2	2000 ohm 1 watt res.
R-3	50,000 ohm 1 watt res.
R-4	50,000 ohm 1 watt res.
R-5	1 meg ohm treble vol. cont.
R-6	2000 ohm $\frac{1}{2}$ watt res.
R-7	2000 ohm $\frac{1}{2}$ watt res.
R-8	500,000 ohm bass tone cont.
R-9	10,000 ohm treble tone cont.
R-10	1.5 meg ohm $\frac{1}{2}$ watt res.
R-11	1 meg ohm $\frac{1}{2}$ watt res.
R-12	1 meg ohm $\frac{1}{2}$ watt res.
R-13	250,000 ohm $\frac{1}{2}$w res.
R-14	100,000 ohm $\frac{1}{2}$w res.
R-15	250,000 ohm $\frac{1}{2}$w res.
R-16	250,000 ohm $\frac{1}{2}$w res.
R-17	500,000 ohm $\frac{1}{2}$w res.
R-18	175 ohm 2 watt res.
R-19	250 ohm 2 watt res.
R-20	10,000 ohm 20w res.
R-21	5000 ohm 1 watt res.
J	Input jack
L	Filter choke: 100 ohm 150ma
T-1	Output trans: 10 watt 2500 ohm to v. c.
T-2	Output trans: 5 watt 5000 ohm to v. c.
T-3	Power trans: 325-0-325 v @ 150ma, 5 v @ 3a, 6.3 v @ 4a
SW	SPST switch
Sockets:	6 octals
Speaker #1	Low freq. PM 12" to 18" cone
Speaker #2	High freq. PM tweeter

6 TUBE 10 WATT 6 VOLT D.C. AMPLIFIER

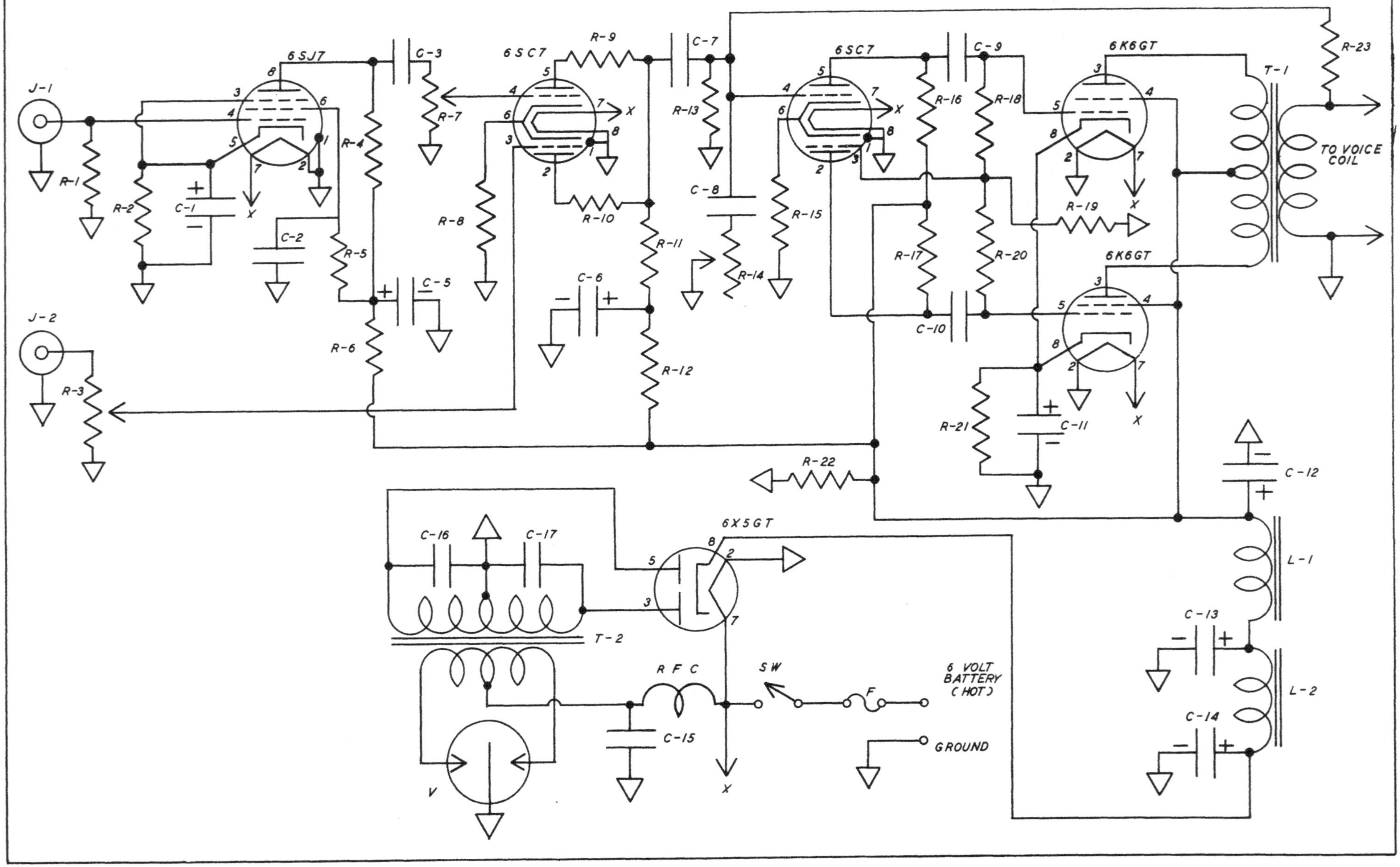

6 TUBE 10 WATT 6 VOLT DC P.A. AMPLIFIER

Shown here is a small 10 watt amplifier suitable for use in an auto or sound truck. A phonograph pickup and microphone can be used simultaneously in this system without interaction. This due to the electronic mixing circuit of the first 6SC7 tube. Volume controls R-3 and R-7 adjust the level of each input circuit, while a tone control R-14, adjusts the amplifier for proper tone. Inverse feedback is applied over two stages of amplification for improved fidelity.

Parts List

C-1	10 mfd 25 v. elec. cond.
C-2	.05 mfd 600 v. paper cond.
C-3	.02 mfd 600 v. paper cond.
C-4	10 mfd 25 v. elec. cond.
C-5	8 mfd 450 v. elec. cond.
C-6	8 mfd 450 v. elec. cond.
C-7	.05 mfd 600 v. paper cond.
C-8	.05 mfd 600 v. paper cond.
C-9	.02 mfd 600 v. paper cond.
C-10	.02 mfd 600 v. paper cond.
C-11	10 mfd 250 v. elec. cond.
C-12	20 mfd 450 v. elec. cond.
C-13	20 mfd 450 v. elec. cond.
C-14	.01 mfd 600 v. paper cond.
C-15	.05 mfd 200 v. paper cond.
C-16	.01 mfd 1600v paper cond.
C-17	.01 mfd 1600v paper cond.
R-1	1 meg ohm ½ watt res.
R-2	1500 ohm ½ watt res.
R-3	500,000 ohm vol. control
R-4	250,000 ohm ½ watt res.
R-5	1 meg ohm ½ watt res.
R-6	50,000 ohm ½ watt res.
R-7	500,000 ohm vol. cont.
R-8	1500 ohm 1 watt res.
R-9	100,000 ohm ½ watt res.
R-10	100,000 ohm ½ watt res.
R-11	200,000 ohm 1 watt res.
R-12	50,000 ohm 1 watt res.
R-13	500,000 ohm ½w res.
R-14	500,000 ohm tone cont.
R-15	1500 ohm 1 watt res.
R-16	100,000 ohm ½w res.
R-17	100,000 ohm ½w res.
R-18	350,000 ohm ½w res.
R-19	150,000 ohm ½w res.
R-20	350,000 ohm ½w res.
R-21	300 ohm 2 watt res.
R-22	25,000 ohm 5w res.
R-23	1 meg ohm ½w res.
V	Non-sync. vib. 4 prong
J-1	Microphone jack
J-2	Phonograph jack
L-1	Filter choke: 200 ohm--75ma
L-2	RF choke: 2.5MH--125MA
T-1	Output trans: 10 watt 10,000 ohm to v.c.
T-2	Power transformer: 250-0-250 v @ 75ma Pri: 6-0-6 v @ 5a
F	10 amp fuse
sw	SPST switch
Sockets:	6 octals 1 - 4 prong
R.F.C.	Choke for vib. "A" supply

6 TUBE 10 WATT A.C. P.A. AMPLIFIER

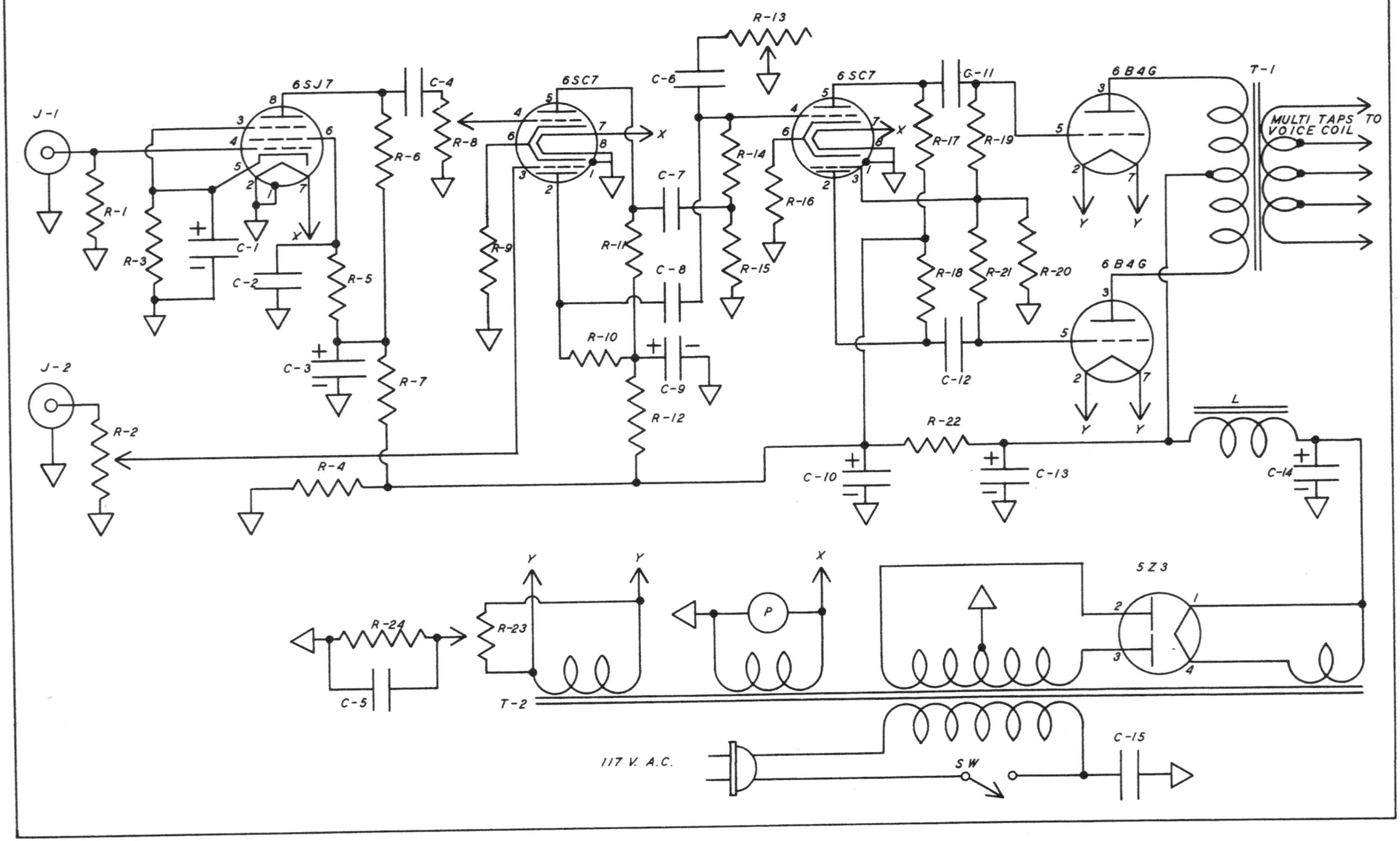

6 TUBE 10 WATT AC P.A. HIGH-FIDELITY AMPLIFIER

This is a 10 watt high-fidelity amplifier which is designed to operate from a high impedance microphone and phonograph pickup simultaneously. A phase inverter tube is used to provide push-pull operation to the power output tubes. The output tubes can be either 6A3 or 6B4G types for six volt operation, or 2A3 tubes if $2\frac{1}{2}$ volts is available for filament operation. A hum balancing control, R-24, can be adjusted to balance out any hum present in the speaker.

The fidelity of the amplifier will depend almost entirely upon the grade of output transformer (T-1) used. Its location on the chassis should be separate from the power transformer and filter choke to keep inductive hum at a minimum.

Parts List

C-1	10 mfd 25 v. elec. cond.
C-2	.1 mfd 600 v. paper cond.
C-3	8 mfd 450 v. elec. cond.
C-4	.02 mfd 600 v. paper cond.
C-5	50 mfd 150 v. elec. cond.
C-6	.01 mfd 600 v. paper cond.
C-7	.05 mfd 600 v. paper cond.
C-8	.05 mfd 600 v. paper cond.
C-9	8 mfd 450 v. elec. cond.
C-10	16 mfd 450 v. elec. cond.
C-11	.1mfd 600 v. paper cond.
C-12	.1 mfd 600 v. paper cond.
C-13	16 mfd 600 v. elec. cond.
C-14	16 mfd 600 v. elec. cond.
C-15	.1 mfd 400 v. paper cond.
R-1	2 meg ohm $\frac{1}{2}$ watt res.
R-2	500,000 ohm vol. control
R-3	1000 ohm 1 watt res.
R-4	25,000 ohm 10 watt res.
R-5	1 meg ohm $\frac{1}{2}$ watt res.
R-6	250,000 ohm $\frac{1}{2}$ watt res.
R-7	50,000 ohm 1 watt res.
R-8	500,000 ohm vol. control
R-9	1500 ohm 1 watt res.
R-10	200,000 ohm $\frac{1}{2}$ watt res.
R-11	200,000 ohm $\frac{1}{2}$ watt res.
R-12	50,000 ohm $\frac{1}{2}$ watt res.
R-13	500,000 ohm tone control
R-14	500,000 ohm $\frac{1}{2}$ watt res.
R-15	500,000 ohm $\frac{1}{2}$ watt res.
R-16	1500 ohm $\frac{1}{2}$ watt res.
R-17	200,000 ohm $\frac{1}{2}$ watt res.
R-18	200,000 ohm $\frac{1}{2}$ watt res.
R-19	250,000 ohm $\frac{1}{2}$ watt res.
R-20	100,000 ohm $\frac{1}{2}$ watt res.
R-21	250,000 ohm $\frac{1}{2}$ watt res.
R-22	7500 ohm 10 watt res.
R-23	50 ohm hum bal. pot. w.w.
R-24	780 ohm 5 watt res.
J-1	Mic. input jack
J-2	Phono. input jack
T-1	Output trans: 15 watt 5000 ohm to multi-tap sec.
T-2	Power transformer: 400-0-400 v @ 125ma 5 v @ 3a 6.3 v @ 1a (x) 6.3 v @ 2a (y)
L	Filter choke: 15Hy - 125ma
SW	SPST toggle switch
P	6.3 v. ind. lamp
Sockets:	5 octals 1 - 4 prong

6 TUBE 12 WATT RECORDING AND PLAYBACK AMPLIFIER

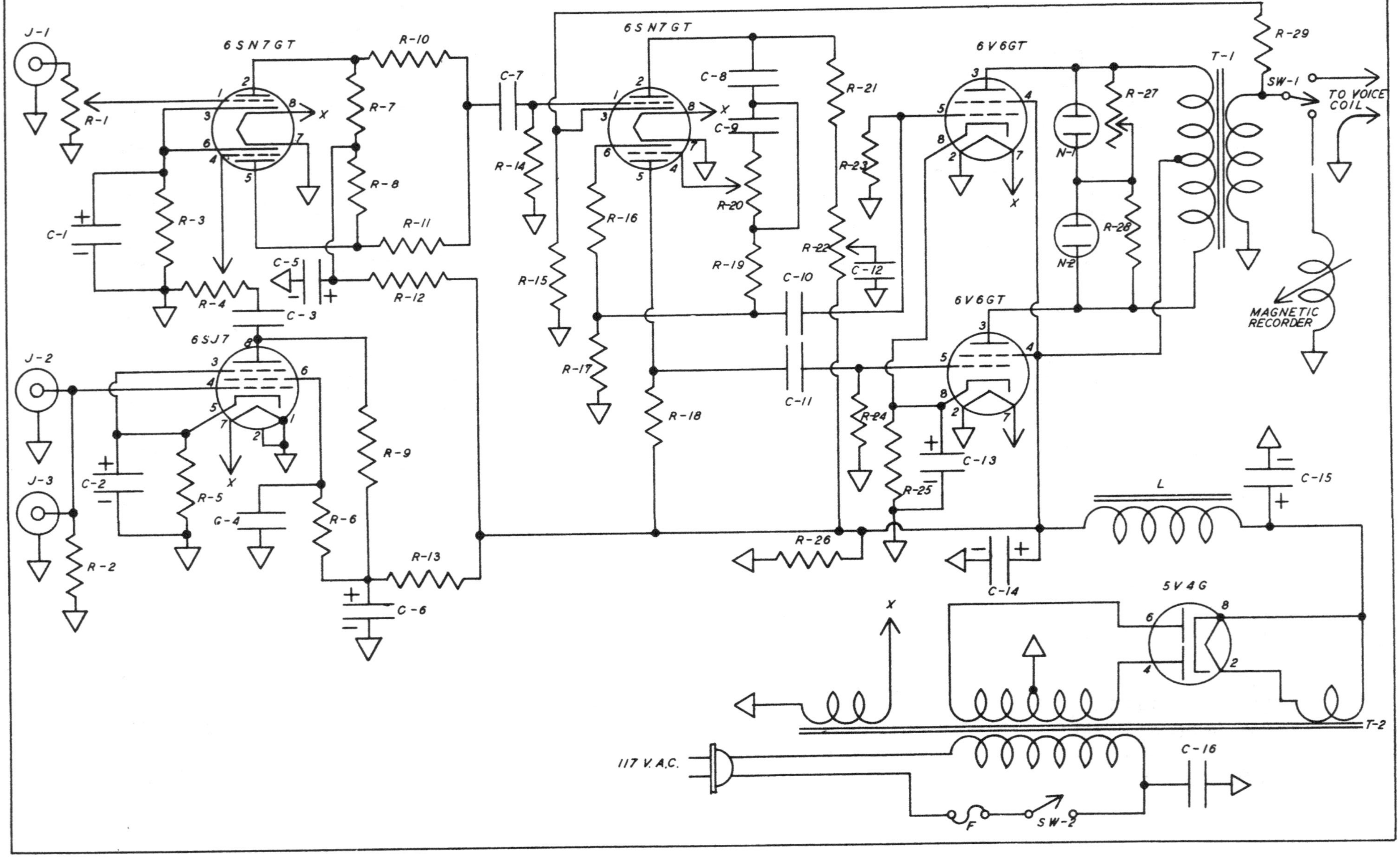

6 TUBE 12 WATT AC RECORDING AND PLAYBACK AMPLIFIER

Shown above is an amplifier suitable for semi-professional recording and playback, having both high gain and output. A phonograph pickup and two similar high impedance microphones can be used in the system; R-1 controls the phonograph volume and R-4 the microphones. This system is designed for an 8 ohm magnetic recording head; sw-1 controls the output for either recording or speaker service. A separate treble, R-20, and bass (R-22) tone control, adjusts the amplifier tone to suit the operator. Two modulation indicators are provided, N-2 for normal modulation, and N-1 for peaks. To adjust the peak indicator, a record is cut and R-27 adjusted so that N-1 flashes just as the record groove is overcut.

Parts List

C-1	25 mfd 25 v. elec. cond.
C-2	25 mfd 25 v. elec. cond.
C-3	.1 mfd 600 v. paper cond.
C-4	.25 mfd 600 v. paper cond.
C-5	8 mfd 450 v. elec. cond.
C-6	8 mfd 450 v. elec. cond.
C-7	.1 mfd 600 v. paper cond.
C-8	.1 mfd 600 v. paper cond.
C-9	.0005 mfd mica cond.
C-10	.1 mfd 600 v. paper cond.
C-11	.1 mfd 600 v. paper cond.
C-12	.5 mfd 400 v. paper cond.
C-13	50 mfd 50 v. elec. cond.
C-14	40 mfd 450 v. elec. cond.
C-15	16 mfd 600 v. elec. cond.
C-16	.1 mfd 400 v. paper cond.
R-1	1 meg ohm vol. control
R-2	2 meg ohm ½ watt res.
R-3	2500 ohm 1 watt res.
R-4	500,000 ohm vol. control
R-5	1000 ohm 1 watt res.
R-6	1 meg ohm ½ watt res.
R-7	40,000 ohm ½ watt res.
R-8	40,000 ohm ½ watt res.
R-9	150,000 ohm ½ watt res.
R-10	20,000 ohm ½ watt res.
R-11	20,000 ohm ½ watt res.
R-12	15,000 ohm 1 watt res.
R-13	50,000 ohm ½ watt res.
R-14	1 meg ohm ½ watt res.
Sockets:	5 octals 1 - 4 prong
R-15	5000 ohm 1 watt res.
R-16	5000 ohm 1 watt res.
R-17	35,000 ohm 1 watt res.
R-18	35,000 ohm 1 watt res.
R-19	750,000 ohm ½ watt res.
R-20	1 meg ohm treble tone cont.
R-21	5000 ohm ½ watt res.
R-22	50,000 ohm bass tone control
R-23	250,000 ohm ½ watt res.
R-24	250,000 ohm ½ watt res.
R-25	250 ohm 3 watt res.
R-26	20,000 ohm 20 watt res.
R-27	250,000 ohm modulation cont.
R-28	1 meg ohm 1 watt res.
R-29	40,000 ohm 1 watt res.
J-1	Phonograph jack
J-2	Microphone jack
J-3	Microphone jack
T-1	Output trans: 12 watt 8000 ohm to 8 ohm v. c.
T-2	Power transformer: 300-0-300 v @ 125ma; 5 v @ 3a; 6.3 v @ 2.5a
L	Filter choke: 100 ohm-- 125ma
M	Magnetic cutter--8 ohm
N-1	¼ watt neon overmod. ind.
N-2	¼ watt neon normal mod. ind.
sw-1	SPDT speaker-recorder switch
sw-2	SPST AC toggle switch

6 TUBE 25 WATT HI-FIDELITY AMPLIFIER

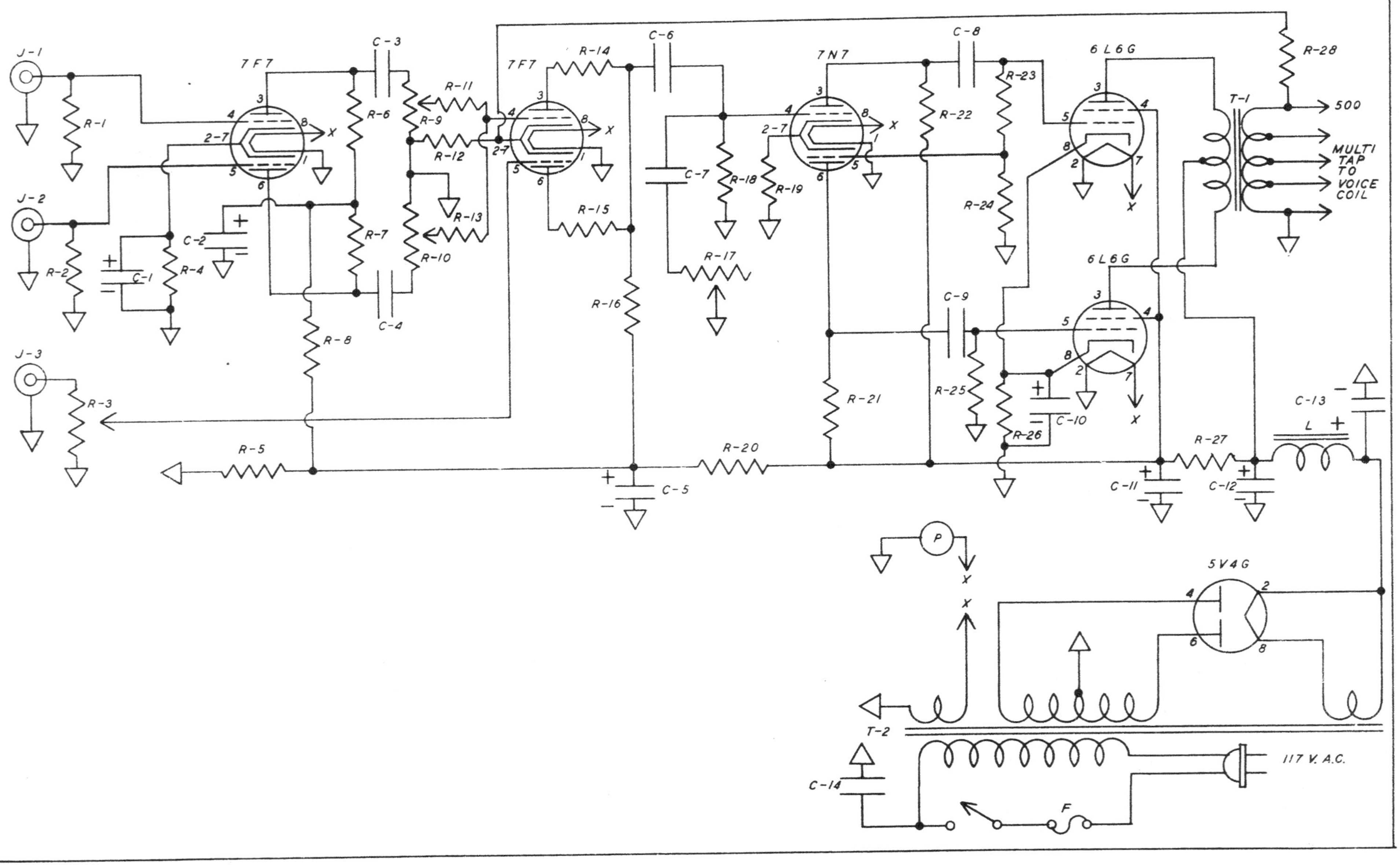

6 TUBE 25 WATT AC HIGH-FIDELITY P.A. AMPLIFIER

This amplifier uses a combination of loctal and octal type tubes for high gain, high output, and compactness. Two microphones and a phonograph pickup, all separately controlled, can be used at will, and with a properly designed output transformer high fidelity results can be realized. Inverse feedback applied over three stages of amplification aids in obtaining this high fidelity response and gives the amplifier smooth operation. A tone control, R-17, can be adjusted to lower the response if desired.

Parts List

C-1	10 mfd 25 v. elec. cond.
C-2	8 mfd 450 v. elec. cond.
C-3	.05 mfd 600 v. paper cond.
C-4	.05 mfd 600 v. paper cond.
C-5	8 mfd 450 v. elec. cond.
C-6	.1 mfd 600 v. paper cond.
C-7	.005 mfd 600 v. paper cond.
C-8	.1 mfd 600 v. paper cond.
C-9	.1 mfd 600 v. paper cond.
C-10	50 mfd 50 v. elec. cond.
C-11	8 mfd 450 v. elec. cond.
C-12	25 mfd 600 v. elec. cond.
C-13	16 mfd 600 v. elec. cond.
C-14	.05 mfd 400 v. paper cond.
R-1	1 meg ohm ½ watt res.
R-2	1 meg ohm ½ watt res.
R-3	500,000 ohm vol. control
R-4	1500 ohm 1 watt res.
R-5	25,000 ohm 10 watt res.
R-6	150,000 ohm ½ watt res.
R-7	150,000 ohm ½ watt res.
R-8	50,000 ohm 1 watt res.
R-9	500,000 ohm volume cont.
R-10	500,000 ohm volume cont.
R-11	100,000 ohm ½ watt res.
R-12	1500 ohm 1 watt res.
R-13	100,000 ohm ½ watt res.
R-14	100,000 ohm 1 watt res.
R-15	100,000 ohm 1 watt res.
R-16	100,000 ohm 1 watt res.
R-17	500,000 ohm tone control
R-18	500,000 ohm ½ watt res.
R-19	2000 ohm 1 watt res.
R-20	5000 ohm 5 watt res.
R-21	50,000 ohm 1 watt res.
R-22	50,000 ohm 1 watt res.
R-23	250,000 ohm ½ watt res.
R-24	15,000 ohm ½ watt res.
R-25	250,000 ohm ½ watt res.
R-26	200 ohm 10 watt res.
R-27	5000 ohm 10 watt res.
R-28	1½ meg ohm ½ watt res.
L	Filter choke: 200 ohm 150ma
T-1	Output trans: 25 watts 6600 ohm to multi-tap sec.
T-2	Power transformer: 350-0-350 v @ 150ma, 5 v @ 3a, 6.3 v @ 4a
J-1	Microphone jack
J-2	Microphone jack
J-3	Phonograph jack
F	3 amp fuse
SW	SPST toggle switch
P	6.3 v. indicator light
Sockets:	3 octal 3 loctal

6 TUBE 45 WATT A.C. FIXED BIAS AMPLIFIER

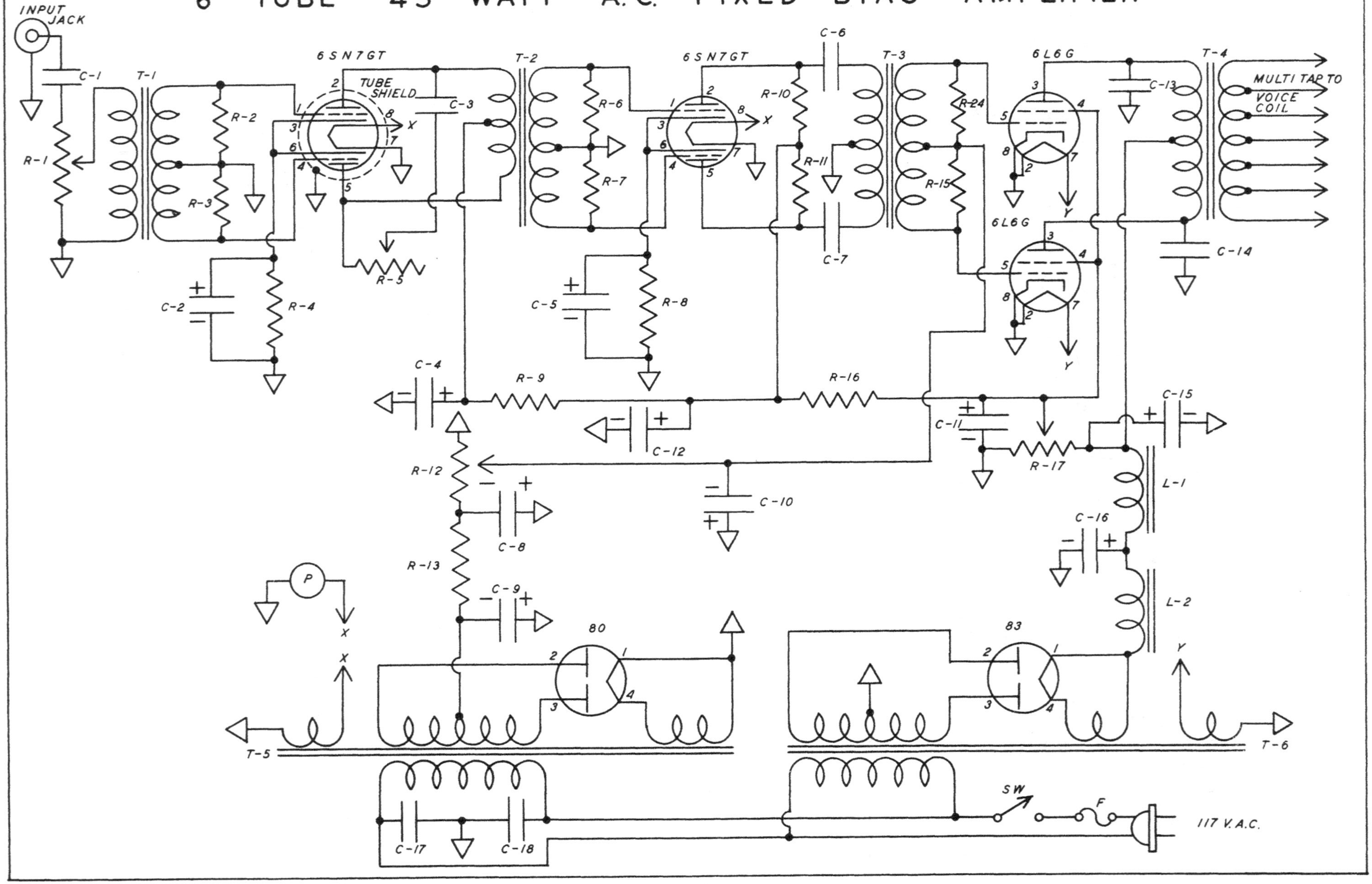

6 TUBE 45 WATT AC FIXED BIAS AMPLIFIER

The above is an all-transformer coupled amplifier which is capable of high power output from any desired input source. Push-pull operation is provided throughout the amplifier, and with good audio frequency transformers, fine response can be obtained. Two rectifiers are used, one for plate supply, the other for bias supply of the 6L6G tubes. In adjusting the circuits, R-12 is adjusted for —$22\frac{1}{2}$ volts and R-17 for plus 275 volts. It is usually advisable to provide a separate power supply chassis from the amplifier chassis to keep hum problems at a minimum. It may also be necessary to stagger the positions of the A.F. transformers for the same reason.

Parts List

C-1	.05 mfd 600 v. paper cond.
C-2	10 mfd 25 v. elec. cond.
C-3	.05 mfd 600 v. paper cond.
C-4	8 mfd 450 v. elec. cond.
C-5	10 mfd 25 v. elec. cond.
C-6	.25 mfd 600 v. paper cond.
C-7	.25 mfd 600 v. paper cond.
C-8	40 mfd 250 v. elec. cond.
C-9	16 mfd 450 v. elec. cond.
C-10	20 mfd 150 v. elec. cond.
C-11	16 mfd 450 v. elec. cond.
C-12	8 mfd 450 v. elec. cond.
C-13	.001 mfd 1000 v. paper cond.
C-14	.001 mfd 1000 v. paper cond.
C-15	16 mfd 600 v. elec. cond.
C-16	16 mfd 600 v. elec. cond.
C-17	.05 mfd 400 v. paper cond.
C-18	.05 mfd 400 v. paper cond.
R-1	500,000 ohm vol. control
R-2	200,000 ohm $\frac{1}{2}$ watt res.
R-3	200,000 ohm $\frac{1}{2}$ watt res.
R-4	750 ohm 1 watt res.
R-5	100,000 ohm tone control
R-6	200,000 ohm $\frac{1}{2}$ watt res.
R-7	200,000 ohm $\frac{1}{2}$ watt res.
R-8	1000 ohm 1 watt res.
R-9	25,000 ohm 1 watt res.
sw	SPST toggle switch
F	3 amp fuse
Sockets:	4 octals 2 - 4 prong
R-10	50,000 ohm 1 watt res.
R-11	50,000 ohm 1 watt res.
R-12	750 ohm 10w semi-adj. res.
R-13	7500 ohm 20 watt res.
R-14	200,000 ohm $\frac{1}{2}$ watt res.
R-15	200,000 ohm $\frac{1}{2}$ watt res.
R-16	2500 ohm 5 watt res.
R-17	10,000 ohm 50 watt semi-adj. res.
L-1	Smoothing choke: 100 ohm - 200ma
L-2	Swinging choke: 3 - 30Hy, 200ma
T-1	AF input trans: input to P.P. grid
T-2	AF interstage P.P. trans: 3:1 ratio
T-3	AF interstage P.P. trans: Class B grids
T-4	Output trans: 50 watts 3800 ohm to multi-tap sec.
T-5	Power trans: 200-0-200 v @ 40ma; 5 v @ 2a; 6.3 v @ 2a
T-6	Power trans: 400-0-400 v @ 200ma; 5 v @ 3a; 6.3 v @ 2a
P	6.3 v indicator lamp

7 TUBE 14 WATT RECORDING AND PLAYBACK AMPLIFIER

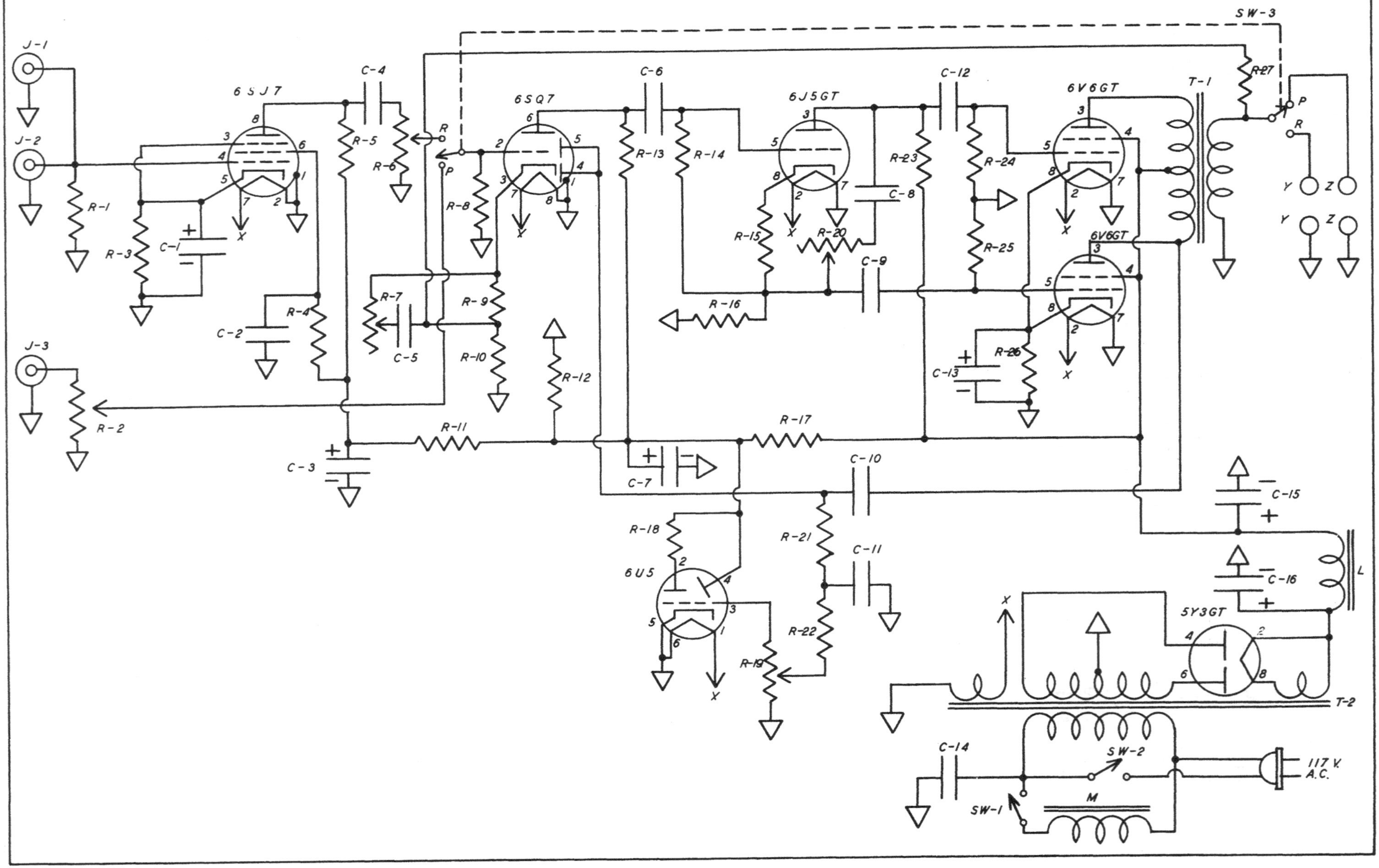

7 TUBE 14 WATT RECORDING AND PLAYBACK AMPLIFIER

Here is a recording-playback amplifier which is somewhat more elaborate than the previous one shown. A recording-playback switch, sw-3, connects the microphone and recorder to the amplifier on "record," and the phonograph and speaker on "playback" positions. A magnetic recorder is again specified. A separate treble and bass tone controls is provided, and a "magic eye" tube is used as a modulation indicator. R-19 is adjusted so that the eye will close just at the point of over-modulation. Two microphone jacks J-1 and J-2 are available if two microphones are to be used. These are controlled by R-6; the phonograph pickup level is set by R-2.

Parts List

C-1	25 mfd 25 v. elec. cond.
C-2	.25 mfd 400 v. paper cond.
C-3	8 mfd 450 v. elec. cond.
C-4	.02 mfd 600 v. paper cond.
C-5	.5 mfd 200 v. paper cond.
C-6	.05 mfd 600 v. paper cond.
C-7	8 mfd 450 v. elec. cond.
C-8	.05 mfd 600 v. paper cond.
C-9	.05 mfd 600 v. paper cond.
C-10	.001 mfd 800 v. paper cond.
C-11	.25 mfd 400 v. paper cond.
C-12	.05 mfd 600 v. paper cond.
C-13	25 mfd 50 v. elec. cond.
C-14	.05 mfd 400 v. paper cond.
C-15	40 mfd 450 v. elec. cond.
C-16	8 mfd 450 v. elec. cond.
R-1	1 meg ohm $\frac{1}{2}$ watt res.
R-2	500,000 ohm vol. control
R-3	1000 ohm 1 watt res.
R-4	1.5 meg ohm $\frac{1}{2}$ watt res.
R-5	250,000 ohm $\frac{1}{2}$ watt res.
R-6	500,000 ohm vol. control
R-7	10,000 ohm treble tone cont.
R-8	2 meg ohm $\frac{1}{2}$ watt res.
R-9	3000 ohm $\frac{1}{2}$ watt res.
R-10	1000 ohm $\frac{1}{2}$ watt res.
R-11	100,000 ohm 1 watt res.
R-12	20,000 ohm 10 watt res.
R-13	250,000 ohm $\frac{1}{2}$ watt res.
R-14	1 meg ohm $\frac{1}{2}$ watt res.
R-15	5000 ohm $\frac{1}{2}$ watt res.
R-16	50,000 ohm $\frac{1}{2}$w res.
R-17	5000 ohm 10w res.
R-18	1 meg ohm $\frac{1}{2}$w res.
R-19	1 meg ohm mod. cont.pot.
R-20	500,000 ohm B. T. cont.
R-21	5 meg ohm $\frac{1}{2}$ watt res.
R-22	5 meg ohm $\frac{1}{2}$ watt res.
R-23	50,000 ohm $\frac{1}{2}$ watt res.
R-24	500,000 ohm $\frac{1}{2}$w res.
R-25	500,000 ohm $\frac{1}{2}$w res.
R-26	200 ohm 5 watt res.
R-27	15,000 ohm 1 watt res.
J-1	Microphone jack
J-2	Microphone jack
J-3	Phonograph jack
Y-Y	8 ohm recording conn.
Z-Z	8 ohm speaker conn.
T-1	Output trans: 15 watt 8000 ohm to 8 ohm
T-2	Power transformer: 325-0-325 v @ 100ma 5 v @ 2a 6.3 v @ 2a
L	Filter choke: 150 ohm-100ma
sw-1	SPST motor switch
sw-2	SPST power switch
sw-3	DPDT record-playback sw.
Sockets:	6 octal 1 - 6 prong
M	Turntable motor

7 TUBE 20 WATT A.C. HI-FIDELITY AMPLIFIER

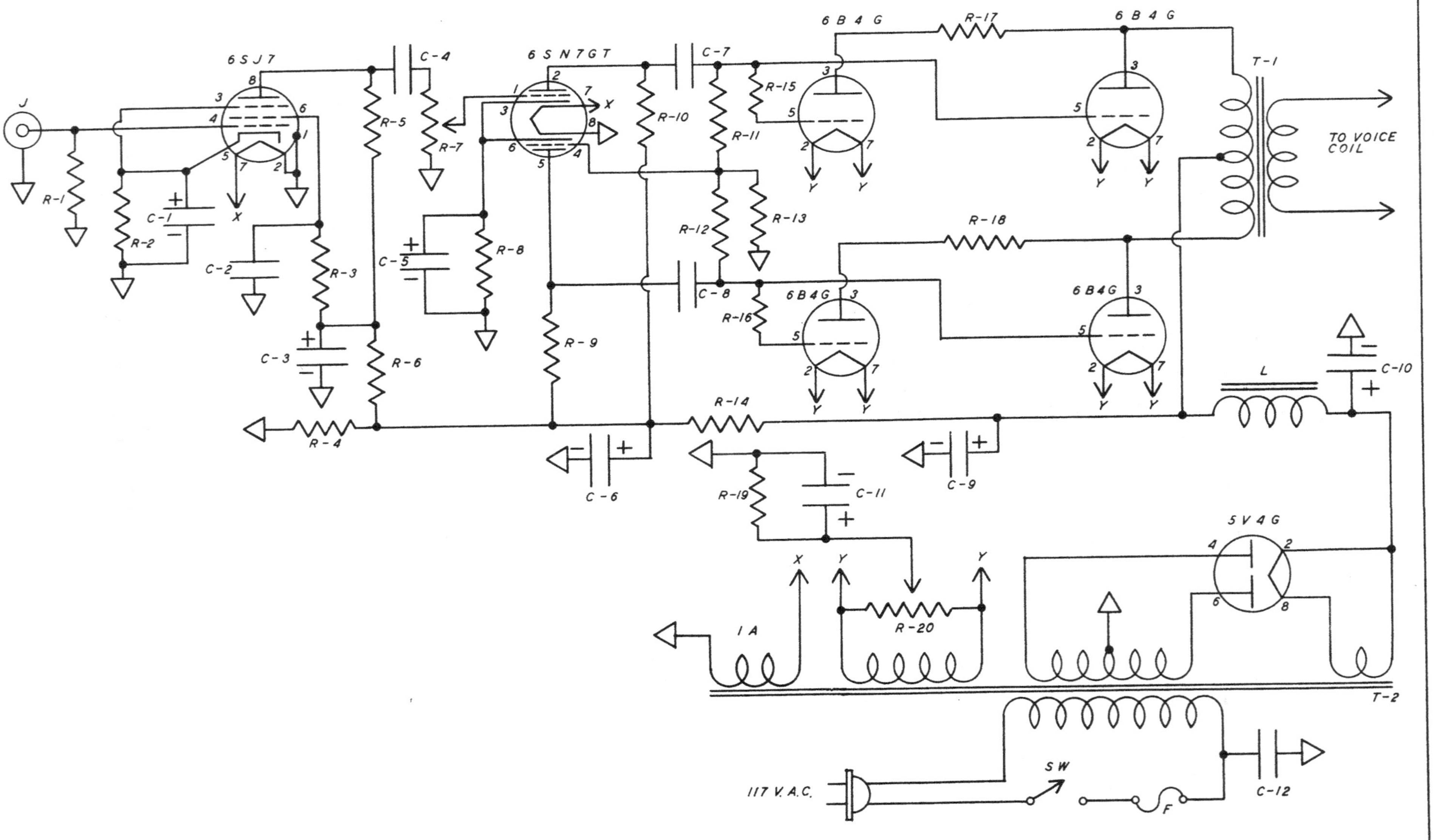

7 TUBE 20 WATT AC HIGH-FIDELITY AMPLIFIER

A phonograph or radio output can be amplified to high levels with this 20 watt high fidelity amplifier. It features push-pull parallel 6B4G tubes in a circuit requiring a very low load impedance. This low impedance, together with a good quality output transformer, will provide high fidelity response, making the amplifier especially suitable for a frequency modulated tuner. R-7 is the volume control, while R-20 balances out any hum present.

Parts List

C-1	25 mfd 25 v. elec. cond.
C-2	.1 mfd 600 v. paper cond.
C-3	8 mfd 450 v. elec. cond.
C-4	.1 mfd 600 v. paper cond.
C-5	25 mfd 25 v. elec. cond.
C-6	8 mfd 450 v. elec. cond.
C-7	.25 mfd 600 v. paper cond.
C-8	.25 mfd 600 v. paper cond.
C-9	40 mfd 450 v. elec. cond.
C-10	16 mfd 600 v. elec. cond.
C-11	20 mfd 150 v. elec. cond.
C-12	.1 mfd 400 v. paper cond.
R-1	500,000 ohm $\frac{1}{2}$ watt res.
R-2	1000 ohm $\frac{1}{2}$ watt res.
R-3	1.5 meg ohm $\frac{1}{2}$ watt res.
R-4	25,000 ohm 10 watt res.
R-5	250,000 ohm $\frac{1}{2}$ watt res.
R-6	25,000 ohm $\frac{1}{2}$ watt res.
R-7	1 meg ohm vol. control
R-8	2500 ohm 1 watt res.
R-9	50,000 ohm 1 watt res.
R-10	50,000 ohm 1 watt res.
R-11	250,000 ohm $\frac{1}{2}$ watt res.
R-12	250,000 ohm $\frac{1}{2}$ watt res.
R-13	100,000 ohm $\frac{1}{2}$ watt res.
R-14	5000 ohm 10 watt res.
R-15	200 ohm $\frac{1}{2}$ watt res.
R-16	200 ohm $\frac{1}{2}$ watt res.
R-17	100 ohm 5 watt res.
R-18	100 ohm 5 watt res.
R-19	400 ohm 5 watt res.
R-20	50 ohm 10w semi-adj. res.
T-1	Output trans: 20 watt 2500 ohm to voice coil
T-2	Power transformer: 375-0-375 v @ 200ma 5 v @ 3a 6.3 v @ 4a 6.3 v @ 1a
L	Filter choke: 100 ohm-200ma
J	Input jack
F	3 amp fuse
sw	SPST toggle switch
Sockets:	7 octals

7 TUBE 45 WATT A.C. P.A. AMPLIFIER

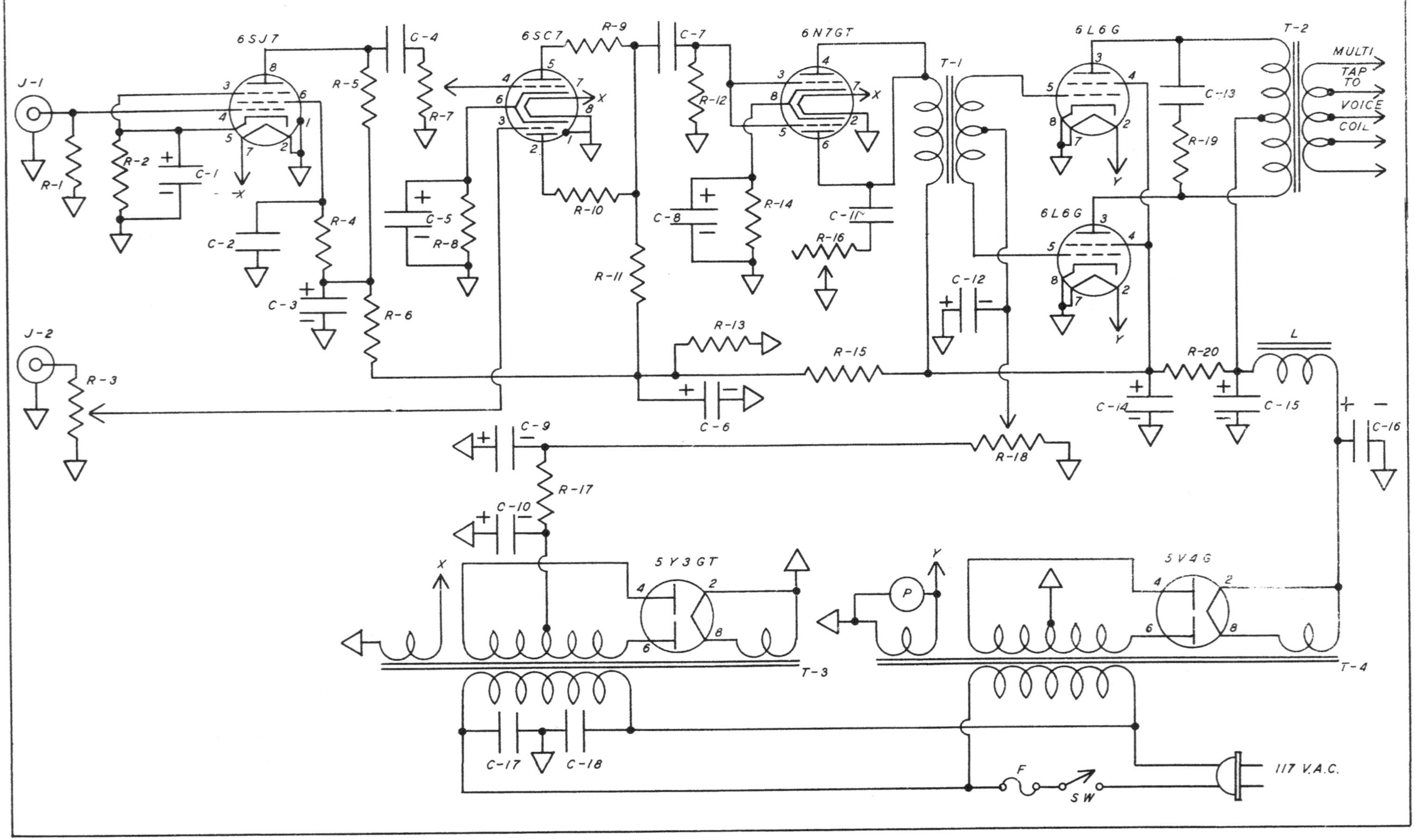

7 TUBE 45 WATT AC P.A. AMPLIFIER

A high power output public address amplifier utilizing push-pull 6L6G tubes is shown. Fixed bias and sufficient driving power for the output tubes provides this high power to the speakers. Voltage dividing resistor (R-18) is adjusted for -22½ volts on the tap. A high impedance microphone and phonograph pickup may be used simultaneously and adjusted separately by means of R-7 and R-3; R-16 is a tone control.

Parts List

C-1	10 mfd 25 v. elec. cond.
C-2	.05 mfd 600 v. paper cond.
C-3	8 mfd 450 v. elec. cond.
C-4	.02 mfd 600 v. paper cond.
C-5	10 mfd 25 v. elec. cond.
C-6	8 mfd 450 v. elec. cond.
C-7	.02 mfd 600 v. paper cond.
C-8	10 mfd 25 v. elec. cond.
C-9	40 mfd 450 v. elec. cond.
C-10	16 mfd 450 v. elec. cond.
C-11	.05 mfd 600 v. paper cond.
C-12	20 mfd 150 v. elec. cond.
C-13	.02 mfd 600 v. paper cond.
C-14	16 mfd 450 v. elec. cond.
C-15	25 mfd 600 v. elec. cond.
C-16	16 mfd 600 v. elec. cond.
C-17	.05 mfd 400 v. paper cond.
C-18	.05 mfd 400 v. paper cond.
R-1	2 meg ohm ½ watt res.
R-2	3000 ohm ½ watt res.
R-3	500,000 ohm vol. control
R-4	500,000 ohm ½ watt res.
R-5	250,000 ohm ½ watt res.
R-6	100,000 ohm ½ watt res.
R-7	500,000 ohm vol. control
R-8	1500 ohm 1 watt res.
R-9	100,000 ohm ½ watt res.
R-10	100,000 ohm ½ watt res.
R-11	100,000 ohm ½w res.
R-12	500,000 ohm ½w res.
R-13	15,000 ohm 10w res.
R-14	1000 ohm 1 watt res.
R-15	5000 ohm 10w res.
R-16	100,000 ohm tone cont.
R-17	7500 ohm 20w res.
R-18	750 ohm 10w semi-adj.
R-19	10,000 ohm 10w res.
R-20	3000 ohm 20w res.
T-1	A.F. class AB input transformer
T-2	Output trans: 50w 3800 ohm to v. c.
T-3	Power transformer: 200-0-200 v @ 40ma 5 v @ 2a 6.3 v @ 2a
T-4	Power transformer: 400-0-400 v @ 200ma 5 v @ 3a 6.3 v @ 2a
L	Filter choke: 100 ohm-200ma
P	6.3 v indicator lamp
J-1	Microphone jack
J-2	Phonograph jack
F	5 amp fuse
SW	SPST toggle switch
Sockets:	7 octals

8 TUBE 12 WATT PHONOGRAPH EXPANDER / COMPRESSOR AMPLIFIER

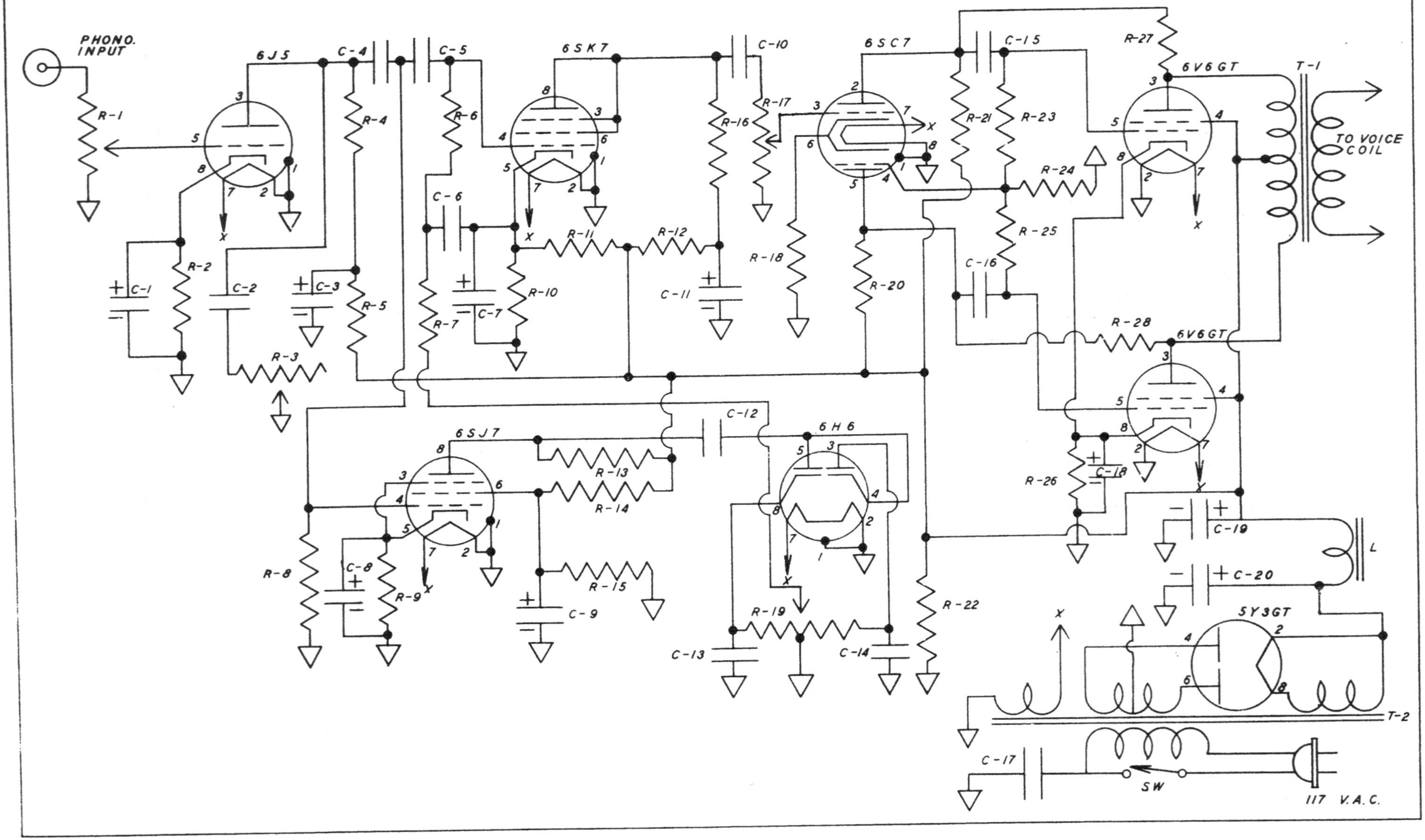

8 TUBE 12 WATT AC PHONO. EXPANDER AND COMPRESSOR AMPLIFIER

This unit is a special phonograph amplifier designed for either expansion or compression of the normal volume level of the recording to any degree. Expansion is sometimes desired to bring out more vividly the heavy passages of a recording than would otherwise appear. Compression is likewise occasionally used to flatten out the response of a recording having too great a variation in volume. Either condition can be obtained by varying R-19; moving the control from the center to the right increases the volume compression, and to the left increases volume expansion. Sufficient gain may be realized in this amplifier to easily reach the 12 watts output at any setting of R-19. NOTE: R-1 should be preset in conjunction with R-19

Parts List

C-1	10 mfd 25 v. elec. cond.
C-2	.05 mfd 600 v. paper cond.
C-3	8 mfd 450 v. elec. cond.
C-4	.1 mfd 600 v. paper cond.
C-5	.1 mfd 400 v. paper cond.
C-6	.5 mfd 200 v. paper cond.
C-7	25 mfd 25 v. elec. cond.
C-8	25 mfd 25 v. elec. cond.
C-9	8 mfd 450 v. elec. cond.
C-10	.05 mfd 600 v. paper cond.
C-11	8 mfd 450 v. elec. cond.
C-12	.05 mfd 600 v. paper cond.
C-13	.25 mfd 200 v. paper cond.
C-14	.25 mfd 200 v. paper cond.
C-15	.05 mfd 600 v. paper cond.
C-16	.05 mfd 600 v. paper cond.
C-17	.1 mfd 400 v. paper cond.
C-18	50 mfd 50 v. elec. cond.
C-19	40 mfd 450 v. elec. cond.
C-20	16 mfd 450 v. elec. cond.
R-1	500,000 ohm input vol. control
R-2	3000 ohm 1 watt res.
R-3	250,000 ohm tone control
R-4	50,000 ohm 1 watt res.
R-5	30,000 ohm 1 watt res.
R-6	750,000 ohm $\frac{1}{2}$ watt res.
R-7	250,000 ohm $\frac{1}{2}$ watt res.
R-8	500,000 ohm $\frac{1}{2}$ watt res.
R-9	2500 ohm 1 watt res.
R-10	3500 ohm 1 watt res.
R-11	50,000 ohm 1 watt res.
R-12	25,000 ohm 1 watt res.
R-13	250,000 ohm 1 watt res.
R-14	100,000 ohm 1 watt res.
R-15	50,000 ohm 1 watt res.
R-16	30,000 ohm 1 watt res.
R-17	500,000 ohm master vol. control
R-18	1500 ohm 1 watt res.
R-19	1 meg ohm pot. cen-tap.
R-20	200,000 ohm $\frac{1}{2}$ watt res.
R-21	200,000 ohm $\frac{1}{2}$ watt res.
R-22	25,000 ohm 10w res.
R-23	500,000 ohm $\frac{1}{2}$w res.
R-24	100,000 ohm $\frac{1}{2}$w res.
R-25	500,000 ohm $\frac{1}{2}$w res.
R-26	200 ohm 5 watt res.
R-27	1 meg ohm $\frac{1}{2}$w res.
R-28	1 meg ohm $\frac{1}{2}$w res.
L	Filter choke: 150 ohm-120ma
T-1	Output trans: 15w 8000 ohm to v. c.
T-2	Power transformer: 300-0-300 v @ 120ma, 5 v @ 2a, 6.3 v @ 3a
sw	SPST switch
Sockets:	8 octals

8 TUBE 25 WATT 4 CHANNEL P. A. AMPLIFIER

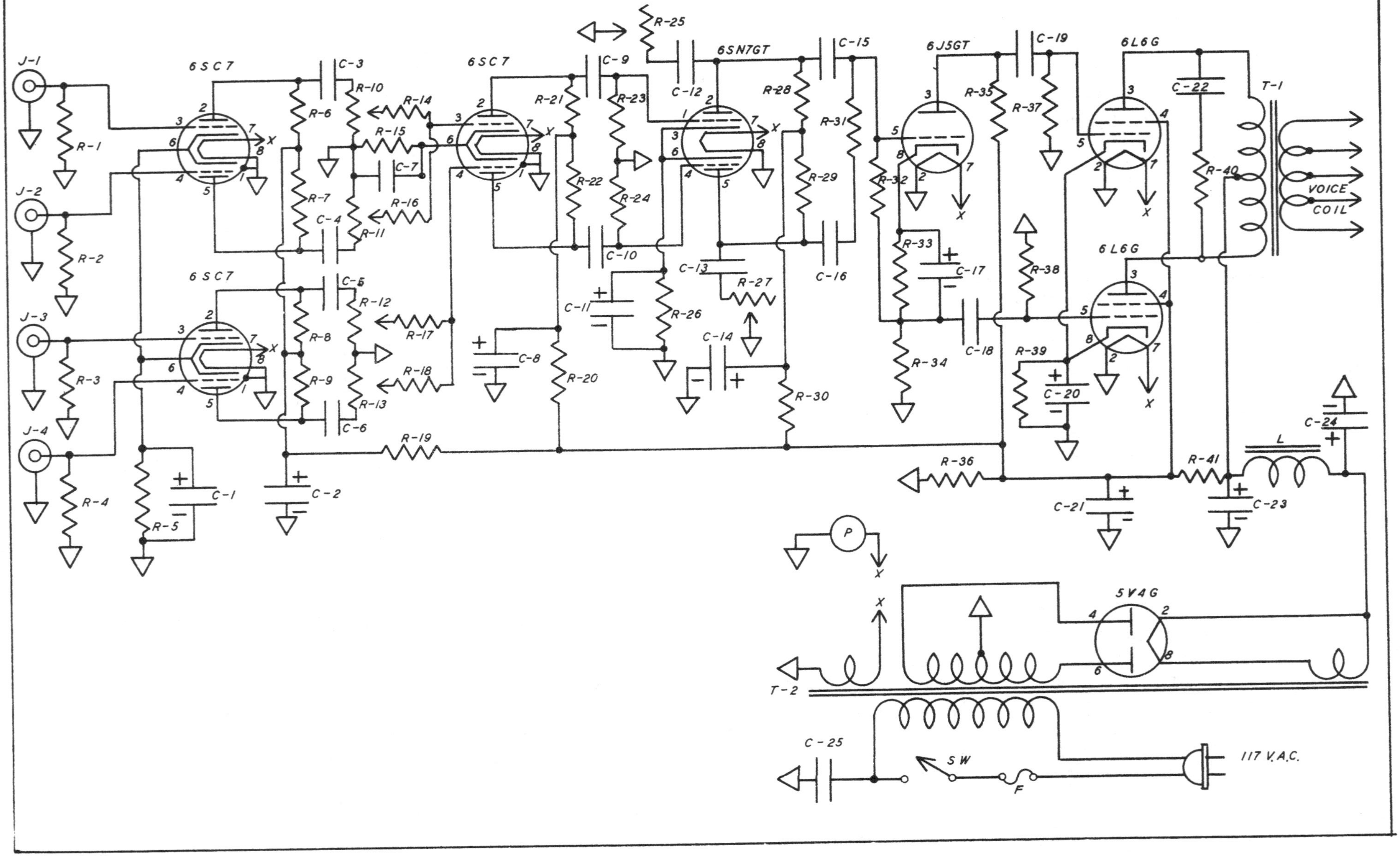

8 TUBE 25 WATT AC 4-CHANNEL P.A. AMPLIFIER

A high power public address amplifier is shown, having two microphone and two phonograph inputs with complete electronic mixing provided on all channels. A separate tone control for the microphone and phonograph channels is used for greater flexibility. High gain and up to 25 watts output is easily obtained. making this a good all-around amplifier. The phono. inputs are designed for very low output cartridges only.

Again a separate chassis for the power supply is desired in constructing the amplifier so as to reduce the interaction between the power supply and input circuits. If this precaution is taken, little trouble should be experienced in getting this amplifier to function properly.

Parts List

C-1	25 mfd 25v elec. cond.	R-3	1 meg ohm ½w res.	R-30	25,000 ohm 1w res.
C-2	8 mfd 450v elec. cond.	R-4	1 meg ohm ½w res.	R-31	100,000 ohm ½w res.
C-3	.05 mfd 600v paper cond.	R-5	750 ohm 1w res.	R-32	500,000 ohm ½w res.
C-4	.05 mfd 600v paper cond.	R-6	200,000 ohm ½w res.	R-33	5000 ohm ½w res.
C-5	.05 mfd 600v.paper cond.	R-7	200,000 ohm ½w res.	R-34	50,000 ohm ½w res.
C-6	.05 mfd 600v paper cond.	R-8	200,000 ohm ½w res.	R-35	50,000 ohm ½w res.
C-7	25 mfd. 25v elec. cond.	R-9	200,000 ohm ½w res.	R-36	15,000 ohm 20w res.
C-8	8 mfd 450v elec. cond.	R-10	500,000 ohm vol. cont.	R-37	500,000 ohm ½w res.
C-9	.1 mfd 600v paper cond.	R-11	500,000 ohm vol. cont.	R-38	500,000 ohm ½w res.
C-10	.1 mfd 600v paper cond.	R-12	500,000 ohm vol. cont.	R-39	250 ohm 10w res.
C-11	25 mfd 25v elec. cond.	R-13	500,000 ohm vol. cont.	R-40	7500 ohm 10w res.
C-12	.05 mfd 600v paper cond.	R-14	1 meg ohm ½w res.	R-41	2000 ohm 20w res.
C-13	.05 mfd 600v paper cond.	R-15	1500 ohm 1w res.	L	Filter choke:
C-14	8 mfd 450v elec. cond.	R-16	1 meg ohm ½w res.		100 ohm-200ma
C-15	.1 mfd 600v paper cond.	R-17	1 meg ohm ½w res.	T-1	Output trans: 25w
C-16	.1 mfd 600v paper cond.	R-18	1 meg ohm ½w res.		9000 ohm to multi-
C-17	10 mfd 25v elec. cond.	R-19	50,000 ohm 1w res.		tap voice coil
C-18	.1 mfd 600v paper cond.	R-20	25,000 ohm 1w res.	T-2	Power transformer:
C-19	.1 mfd 600v paper cond.	R-21	50,000 ohm ½w res.		375-0-375v @ 200ma
C-20	50 mfd 50v elec. cond.	R-22	50,000 ohm ½w res.		5v @ 3a
C-21	16 mfd 450v elec. cond.	R-23	1 meg ohm ½w res.		6.3v @ 4a
C-22	.01 mfd 1000v paper cond.	R-24	1 meg ohm ½w res.	J-1	Microphone jack
C-23	25 mfd 600v elec. cond.	R-25	100,000 ohm tone cont.	J-2	Microphone jack
C-24	16 mfd 600v elec. cond.	R-26	3000 ohm 1w res.	J-3	Phonograph jack
C-25	.1 mfd 400v paper cond.	R-27	100,000 ohm tone cont.	J-4	Phonograph jack
R-1	1 meg ohm ½ watt res.	R-28	50,000 ohm ½w res.	P	6.3v ind. lamp
R-2	1 meg ohm ½ watt res.	R-29	50,000 ohm ½w res.	F	5 amp fuse
				SW	SPST switch
				Sockets:	8 octals

8 TUBE 50 WATT TRANSMITTER MODULATOR

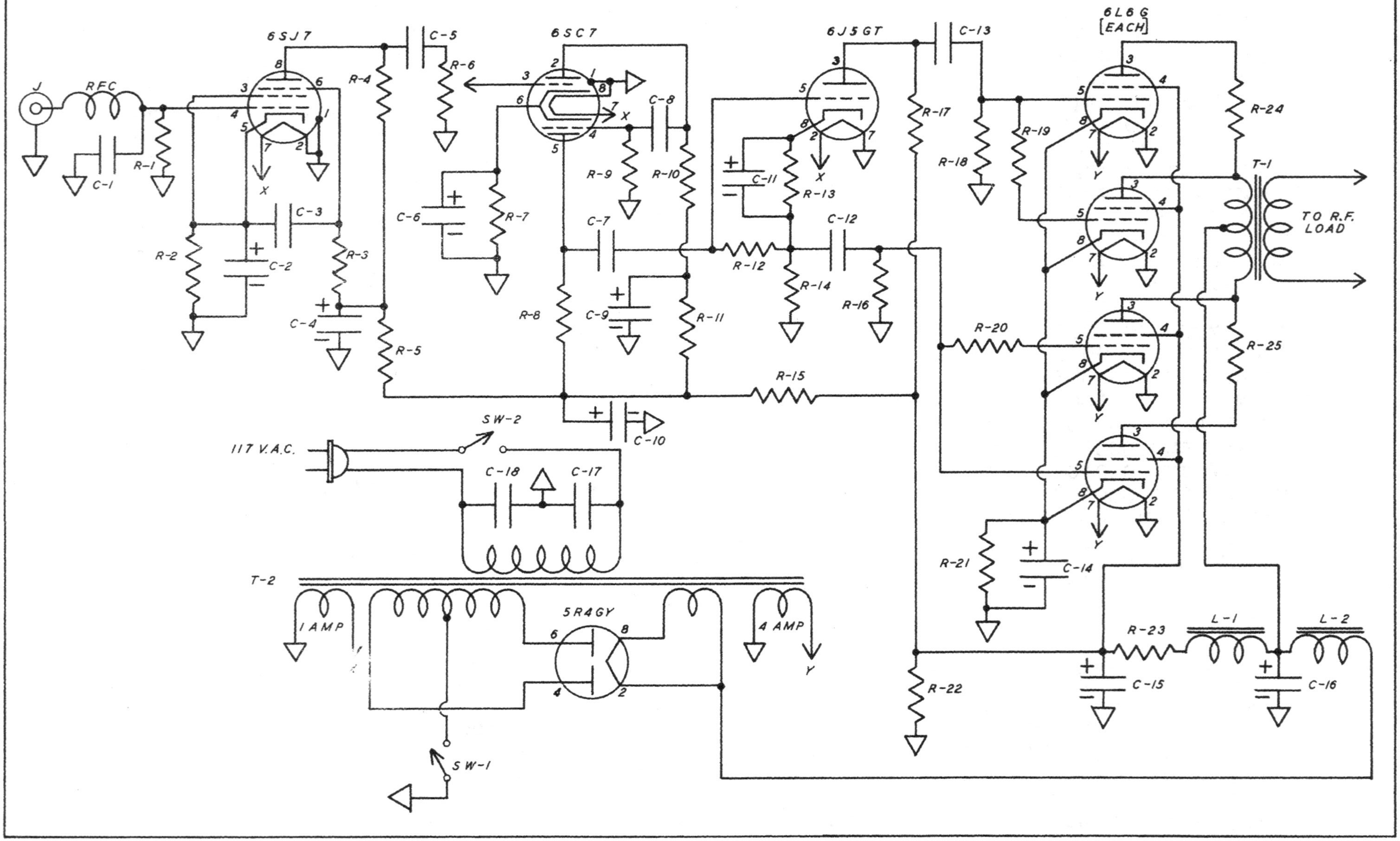

8 TUBE 50 WATT AC TRANSMITTER MODULATOR

Here is a 50 watt modulator suitable for amateur radio service which uses fairly low voltage type tubes--considering the power obtained. No audio driver transformer is needed for the push-pull parallel 6L6G tubes, thereby reducing the size and cost of the unit. A single high impedance microphone input is shown which is the usual practice for an amateur radio station. If it is desired to use two microphones, another jack could be connected parallel to the input jack (J-1). Sw-1 is used to turn the modulator on during transmissions; it may be replaced by a relay if desired.

Parts List

C-1	.001 mfd 200v mica cond.
C-2	10 mfd 25v elec. cond.
C-3	.05 mfd 600v paper cond.
C-4	8 mfd 450v elec. cond.
C-5	.01 mfd 600v paper cond.
C-6	10 mfd 25v elec. cond.
C-7	.01 mfd 600v paper cond.
C-8	.01 mfd 600v paper cond.
C-9	8 mfd 450v elec. cond.
C-10	8 mfd 450v elec. cond.
C-11	10 mfd 25v elec. cond.
C-12	.01 mfd 600v paper cond.
C-13	.01 mfd 600v paper cond.
C-14	10 mfd 50v elec. cond.
C-15	16 mfd 450v elec. cond.
C-16	25 mfd 600v elec. cond.
C-17	.05 mfd 400v paper cond.
C-18	.05 mfd 400v paper cond.
R-1	2 meg ohm ½ watt res.
R-2	1500 ohm ½ watt res.
R-3	500,000 ohm ½ watt res.
R-4	250,000 ohm ½ watt res.
R-5	100,000 ohm ½ watt res.
R-6	500,000 ohm vol. cont.
R-7	1500 ohm 1 watt res.
R-8	100,000 ohm ½ watt res.
R-9	500,000 ohm ½ watt res.
R-10	100,000 ohm ½ watt res.
R-11	25,000 ohm ½ watt res.
R-12	500,000 ohm ½ watt res.
R-13	5000 ohm ½ watt res.
R-14	50,000 ohm ½ watt res.
R-15	10,000 ohm 1 watt res.
R-16	500,000 ohm ½ watt res.
R-17	50,000 ohm ½ watt res.
R-18	500,000 ohm ½ watt res.
R-19	200 ohm ½ watt res.
R-20	200 ohm ½ watt res.
R-21	100 ohm 10 watt res.
R-22	20,000 ohm 20 watt res.
R-23	2000 ohm 20 watt res.
R-24	100 ohm 5 watt res.
R-25	100 ohm 5 watt res.
L-1	Filter choke: 100 ohm-100ma
L-2	Filter choke: 3-30Hy, 300ma
T-1	Output trans: 50 watt 4500 ohm to multi-tap secondary
T-2	Power transformer: 400-0-400v @ 300ma 5v @ 3a 6.3v @ 4a 6.3v @ 1a
sw-1	SPST communication sw
sw-2	SPST toggle switch
J	Input microphone jack
RFC	2½MH radio freq. choke
Sockets:	8 octals

9 TUBE 45 WATT A.C. P.A. AMPLIFIER

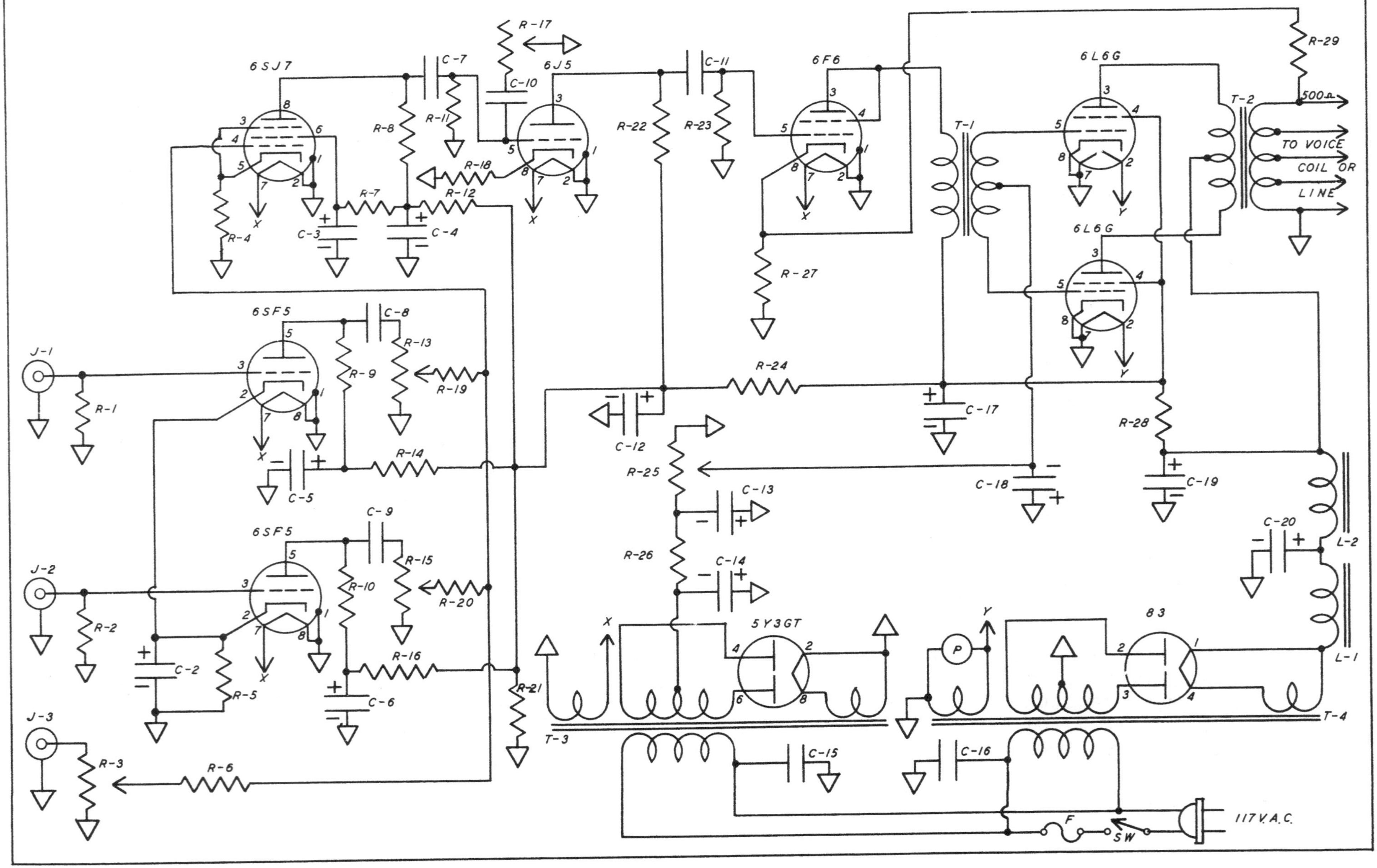

9 TUBE 45 WATT AC P.A. AMPLIFIER

A 45 watt public address amplifier is shown having two microphone and one phonograph inputs. The amplifier uses fixed bias in its output circuit for greater power, and R-25 is pre-adjusted for $-22\frac{1}{2}$ volts bias on the 6L6G tubes. An inverse feedback network consisting of R-29 is incorporated for improved tone quality.

Parts List

C-1	25 mfd 25v elec. cond.
C-2	25 mfd 25v elec. cond.
C-3	.05 mfd 600v paper cond.
C-4	8 mfd 450v elec. cond.
C-5	8 mfd 450v elec. cond.
C-6	8 mfd 450v elec. cond.
C-7	.05 mfd 600v paper cond.
C-8	.05 mfd 600v paper cond.
C-9	.05 mfd 600v paper cond.
C-10	.005 mfd 400v paper cond.
C-11	.1 mfd 600v paper cond.
C-12	8 mfd 450v elec. cond.
C-13	20 mfd 250v elec. cond.
C-14	20 mfd 450v elec. cond.
C-15	.05 mfd 400v paper cond.
C-16	.05 mfd 400v paper cond.
C-17	16 mfd 450v elec. cond.
C-18	20 mfd 150v elec. cond.
C-19	40 mfd 600v elec. cond.
C-20	16 mfd 600v elec. cond.
R-1	1 meg ohm $\frac{1}{2}$ watt res.
R-2	1 meg ohm $\frac{1}{2}$ watt res.
R-3	500,000 ohm vol. cont.
R-4	1500 ohm $\frac{1}{2}$ watt res.
R-5	2000 ohm 1 watt res.
R-6	250,000 ohm $\frac{1}{2}$ watt res.
R-7	1 meg ohm $\frac{1}{2}$ watt res.
R-8	250,000 ohm $\frac{1}{2}$ watt res.
R-9	250,000 ohm $\frac{1}{2}$ watt res.
R-10	250,000 ohm $\frac{1}{2}$ watt res.
R-11	1 meg ohm $\frac{1}{2}$ watt res.
R-12	50,000 ohm 1 watt res.
R-13	500,000 ohm vol. cont.
R-14	50,000 ohm $\frac{1}{2}$ watt res.
R-15	500,000 ohm vol. control
R-16	50,000 ohm $\frac{1}{2}$ watt res.
R-17	500,000 ohm vol. control
R-18	4000 ohm 1 watt res.
R-19	250,000 ohm $\frac{1}{2}$ watt res.
R-20	250,000 ohm $\frac{1}{2}$ watt res.
R-21	20,000 ohm 10 watt res.
R-22	75,000 ohm 1 watt res.
R-23	500,000 ohm $\frac{1}{2}$ watt res.
R-24	2000 ohm 5 watt res.
R-25	2000 ohm 10w adj. res.
R-26	10,000 ohm 10w res.
R-27	750 ohm 5 watt res.
R-28	3500 ohm 20 watt res.
R-29	250,000 ohm 1 watt res.
L-1	Filter swinging choke-200ma
L-2	Filter smoothing choke-200ma
T-1	Class AB2 input trans.
T-2	Output trans: 50 watt 3800 ohm to multi-tap
T-3	Power trans: Bias supply 200-0-200 v @ 40ma 5 v @ 2a 6.3 v @ 2a
T-4	Power trans: power supply 375-0-375 v @ 200ma 5 v @ 3a 6.3 v @ 2a
J-1	Microphone jack
J-2	Microphone jack
J-3	Phonograph jack
F	5 amp fuse
SW	SPST switch
P	6v indicator lamp
Sockets:	8 octals 1-4 prong

10 TUBE 15 WATT COMPACT A.C. P.A. AMPLIFIER

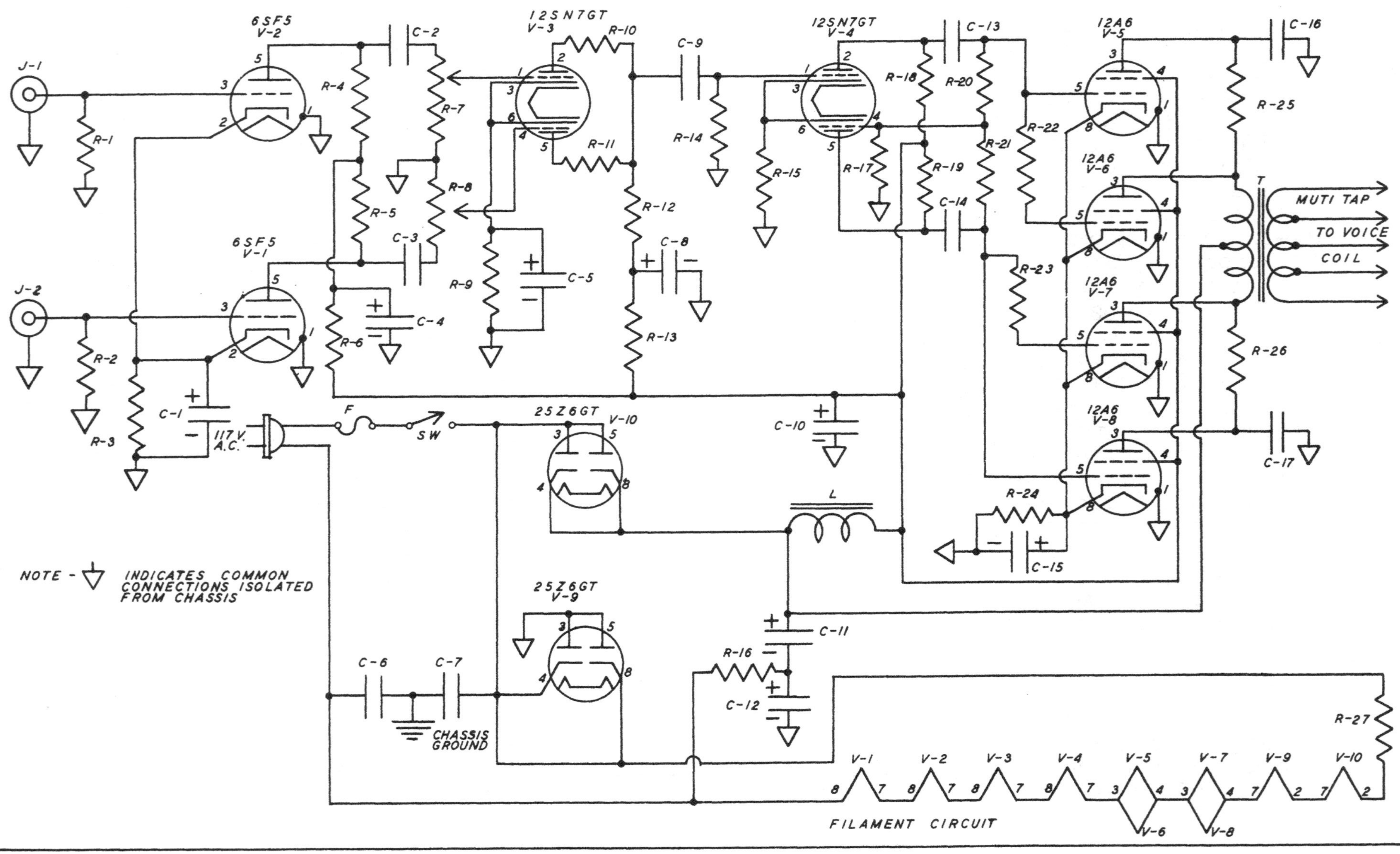

10 TUBE 15 WATT COMPACT AC P.A. AMPLIFIER

Here is a special "transformerless" amplifier having a series filament circuit and a voltage doubler rectifier capable of 15 watts output. It has sufficient gain for high impedance microphone operation, and is designed to handle two of them through an electronic mixing circuit. A low level phonograph pickup may be used in one channel if desired.

Hum is usually the most difficult problem to overcome in this type of amplifier, but if reasonable care is exercised in placing components, and if good parts are used, plus provision for efficient bonding, good results should be obtained.

Parts List

C-1	25 mfd 25v elec. cond.
C-2	.05 mfd 600v. paper cond.
C-3	.05 mfd 600v paper cond.
C-4	10 mfd 250v elec. cond.
C-5	25 mfd 25v elec. cond.
C-6	.05 mfd 400v paper cond.
C-7	.05 mfd 400v paper cond.
C-8	10 mfd 250v elec. cond.
C-9	.05 mfd 600v paper cond.
C-10	50 mfd 250v elec. cond.
C-11	30 mfd 250v elec. cond.
C-12	30 mfd 250v elec. cond.
C-13	.05 mfd 600v paper cond..
C-14	.05 mfd 600v paper cond.
C-15	25 mfd 25v elec. cond.
C-16	.005 mfd 600v paper cond.
C-17	.005 mfd 600v paper cond.
R-1	1 meg ohm ½ watt res.
R-2	1 meg ohm ½ watt res.
R-3	2000 ohm 1 watt res.
R-4	250,000 ohm ½ watt res.
R-5	250,000 ohm ½ watt res.
R-6	50,000 ohm 1 watt res.
R-7	500,000 ohm vol. control
R-8	500,000 ohm vol. control
R-9	2000 ohm 1 watt res.
R-10	200,000 ohm 1 watt res.
R-11	200,000 ohm 1 watt res.
R-12	200,000 ohm 1 watt res.
R-13	50,000 ohm 1 watt res.
R-14	500,000 ohm ½ watt res.
R-15	2000 ohm 1 watt res.
R-16	50 ohm 1 watt res.
R-17	150,000 ohm ½ watt res.
R-18	150,000 ohm ½ watt res.
R-19	150,000 ohm ½ watt res.
R-20	300,000 ohm ½ watt res.
R-21	300,000 ohm ½ watt res.
R-22	100 ohm ½ watt res.
R-23	100 ohm ½ watt res.
R-24	100 ohm 10 watt res.
R-25	50 ohm 1 watt res.
R-26	50 ohm 1 watt res.
R-27	10 ohm 10 watt res.
L	Filter choke: 100ohm-150ma
T	Output transformer: 15 watt 5000 ohm to multi-tap voice coil
J-1	Microphone or phono. jack
J-2	" " " "
F	5 amp fuse
sw	SPST switch
Sockets:	10 octals

11 TUBE 75 WATT A.C. P.A. AMPLIFIER

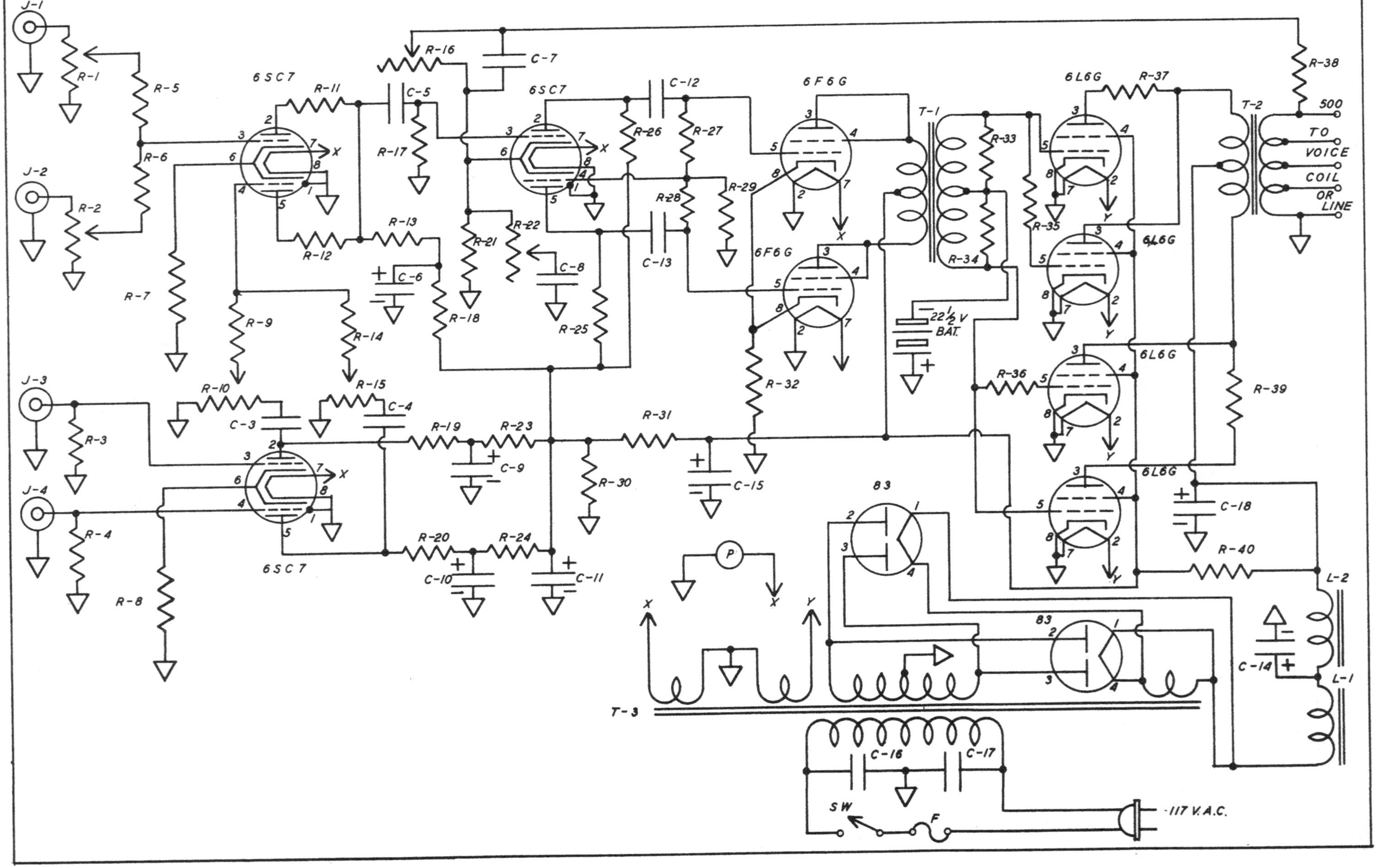

11 TUBE 75 WATT AC P.A. AMPLIFIER

This last amplifier is a high power, high gain unit, having two microphone and two phonograph inputs, separate bass and treble tone controls, inverse feedback, and a fixed bias push-pull parallel output circuit. A $22\frac{1}{2}$ volt battery is used for bias purposes rather than a bias power supply. A battery so used will give long service. A novelity in the tone control circuits is the location of the controls. Being in the inverse feedback circuit, the controls serve to attenuate the amount of feedback at some frequencies more than others, giving tone control effects.

It is best to mount the power supply on a separate chassis from the amplifier so as to eliminate all possibilities of coupling between them. It is advisable to make periodic checks on the bias battery to make certain a full $22\frac{1}{2}$ volts is available, otherwise short life for the power output tubes can be expected.

Parts List

C-3	.05 mfd 600v paper cond.	R-11	100,000 ohm $\frac{1}{2}$w res.	R-37	100 ohm 5w res.
C-4	.05 mfd 600v paper cond.	R-12	100,000 ohm $\frac{1}{2}$w res.	R-38	1.5 meg ohm 1w res.
C-5	.05 mfd 600v paper cond.	R-13	100,000 ohm $\frac{1}{2}$w res.	R-39	100 ohm 5w res.
C-6	8 mfd 450v elec. cond.	R-14	250,000 ohm $\frac{1}{2}$w res.	R-40	2000 ohm 25w res.
C-7	.003 mfd 600v paper cond.	R-15	500,000 ohm v. cont.	L-1	Filter choke: swing-
C-8	.25 mfd 600v paper cond.	R-16	5 meg ohm B.T. cont.		ing - 375ma
C-9	8 mfd 450v elec. cond.	R-17	500,000 ohm $\frac{1}{2}$w res.	L-2	Filter choke: smooth-
C-10	8 mfd 450v elec. cond.	R-18	50,000 ohm $\frac{1}{2}$w res.		ing - 375ma
C-11	8 mfd 450v elec. cond.	R-19	200,000 ohm $\frac{1}{2}$w res.	T-1	Class AB Input trans.
C-12	.1 mfd 600v paper cond.	R-20	200,000 ohm $\frac{1}{2}$w res.	T-2	Output trans: 75 watt
C-13	.1 mfd 600v paper cond.	R-21	1500 ohm 2w res.		2000 ohm to multi-tap
C-14	16 mfd 600v elec. cond.	R-22	10,000 ohm T.T. cont.		secondary
C-15	8 mfd 450v elec. cond.	R-23	50,000 ohm $\frac{1}{2}$w res.	T-3	Power transformer:
C-16	.05 mfd 400v paper cond.	R-24	50,000 ohm $\frac{1}{2}$w res.		400-0-400 v @ 375ma
C-17	.05 mfd 400v paper cond.	R-25	50,000 ohm $\frac{1}{2}$w res.		5 v @ 6a
C-18	25 mfd 600v elec. cond.	R-26	50,000 ohm $\frac{1}{2}$w res.		6.3 v @ 3a
R-1	500,000 ohm vol. cont.	R-27	500,000 ohm $\frac{1}{2}$w res.		6.3 v @ 4a
R-2	500,000 ohm vol. cont.	R-28	100,000 ohm $\frac{1}{2}$w res.	J-1	Phono. input jack
R-3	1 meg ohm $\frac{1}{2}$ watt res.	R-29	500,000 ohm $\frac{1}{2}$w res.	J-2	Phono. input jack
R-4	1 meg ohm $\frac{1}{2}$ watt res.	R-30	10,000 ohm 20w res.	J-3	Microphone input jack
R-5	250,000 ohm $\frac{1}{2}$ watt res.	R-31	10,000 ohm 20w res.	J-4	Microphone input jack
R-6	250,000 ohm $\frac{1}{2}$ watt res.	R-32	300 ohm 2w res.	F	5 amp fuse
R-7	1500 ohm 1 watt res.	R-33	500,000 ohm $\frac{1}{2}$w res.	sw	SPST toggle switch
R-8	1500 ohm 1 watt res.	R-34	500,000 ohm $\frac{1}{2}$w res.	P	6.3v indicator lamp
R-9	250,000 ohm $\frac{1}{2}$ watt res.	R-35	200 ohm $\frac{1}{2}$w res.	Sockets:	9 octals, 2-4 prong
R-10	500,000 ohm vol. cont.	R-36	200 ohm $\frac{1}{2}$w res.	Battery:	$22\frac{1}{2}$v dry battery

SERVICING YOUR AMPLIFIER

Having completed your amplifier, it is hoped that everything is working to your complete satisfaction. However, in the event you are having difficulty, the following discussion may be of some assistance.

Generally speaking, the improper operation of audio amplifiers may be attributed to the following:

I. Unequal output voltage over the audio frequency range. For practical purposes this may be considered to be the frequency range from 50 to 12,000 cycles. Although the complete audio range may extend from about 20 to 20,000 cycles, the cost of satisfactorily reproducing such a band of frequencies would be prohibitive. As a matter of fact, even the finest of theatre sound systems generally operate within a band from 50 to 8500 or 9000 cycles. The greatest cost of High-Fidelity reproduction is in the acoustic system which includes the loudspeaker, frequency dividing networks, (if any), and an effective distributor of sound, consisting of horn or baffle, or a combination of both. So if we limit ourselves to good frequency reproduction in the range from 50 to 12,000 cycles, we will have excellent fidelity when considered from all standpoints.

Suppose we have built an amplifier from a circuit which specifications were designed for wide frequency response, but checking it with an output meter and an audio oscillator, we discover an un-satisfactory frequency range. We would then have to take into consideration the following:

a. Defective tubes.

b. Improper layout of parts.

c. Defective components, or parts not designed for wide range frequency response.

d. Insufficient mechanical rigidity.

e. Poorly soldered connections.

f. Tone controls improperly set.

g. Improper match of input or output impedances.

Any one of the above could be the source of trouble, for circuit efficiency is very important in obtaining good response. In a resistance coupled amplifier the most critical part is generally the output transformer---something which the constructor occasionally obtains without much forethought! So watch item "c"! Item "g", although external to the amplifier proper, may greatly affect the response, vis: matching an 8 ohm voice coil of a speaker to the 2 ohm tap on the output transformer. or matching a high impedance crystal phonograph pickup to the input of an amplifier designed to work from a carbon microphone, will both result in poor operation.

II. Some causes of hum.

a. Defective components, generally tubes, transformers or leaky condensers.

b. Improper layout of parts.

Running high impedance grid or plate circuits close to

each other or close to an A.C. field is a common trouble. Inductive coupling between the power transformer, filter chokes, output transformer or audio transformers if used, or inductive loops within the chassis itself, may be a prolific source of trouble. It is best to run all ground connections to one common point on the chassis by means of a heavy wire. In some cases this point may have to be shifted to different parts of the chassis, but generally grounding near the cathode of the input tube is quite satisfactory. It is also advisable to keep all A.C. wires close to the chassis and run them along the corners if possible. Make certain these wires are sufficiently rigid and large enough in diameter to carry the required current. For the usual amplifier, #18 gauge wire is satisfactory for filament circuits. Incidentally, twisted A.C. wires will give less trouble than parallel run wires. When mounting the transformers on the chassis, keep them at right angles to each other and away from any high impedance source such as grid and plate leads and parts associated with them. Where necessary, shield thoroughly.

In one instance we experienced, all the above precautions were observed but audible hum still persisted. It was finally eliminated by turning the output transformer at a critical angle of forty-five degrees. Later, the power transformer was found to have high inductive leakage which necessitated this action.

c. Hum pickup from an outside source.

In a well shielded amplifier outside inductive pickup by the component parts may be disregarded. By-passing the 115 volt A.C. wires near the power transformer with a .1 mfd or .25 mfd condenser connected to the chassis, helps in eliminating a potential source of noise and hum. If hum is being fed into the input connections of the amplifier, good grounding and shielding of these input leads to the grid of the first amplifier tube will generally suffice.

d. Insufficient filtering.

Normally, the most common cause of hum is in the filtered "B" supply. Here the remedy should be obvious; increase the capacity of condensers or inductance of filter chokes, or both if necessary. Occasionally two or three section filter condensers are used in an amplifier that may be partially shorted between sections. This may cause a hum which can only be corrected by replacing the entire condenser unit. Hum sometimes originates through the common plate supply circuits, between high gain or low level amplifier stages. This can be corrected by using decoupling resistors and condensers in the plate return circuits of the offending stages.

e. Other sources of hum.

A listing of the many other possibilities of hum should include; open grid circuits, defective phonograph motor, defective hum bucking coil on a speaker, shorted turns on a speaker field coil, too low capacity on cathode-to-ground circuits, insufficient shielding of high impedance circuits, filament return of directly heated tubes not electrically center tapped, poorly soldered connections, use of bass tone control which resonates at either 60 or 120 cycles, hum pickup from tone control choke when used. Where very high gain amplifiers are involved it is best to mount the power supply on a seperate chassis and connect both to a good ground by a common bond.

III. Amplitude distortion in the amplifier. This due to:

a. Overloading of the tubes with an excessive signal.

b. Operation on the non-linear portion of the grid voltage plate current characteristics of a tube, i.e., incorrect plate, grid, filament or screen voltages.

c. Improper impedance matching of input or output circuit.

d. Defective audio transformer where used.

e. Unequal voltages on the grids of a push pull amplifier.

f. Using output tubes of insufficient capacity to handle the power output required.

h. Gassy or otherwise defective tubes.

i. Unbalanced audio transformers when used in push pull circuits.

IV. Inverse feedback problems.

As you will note, most of the amplifier diagrams in this manual use some form of feedback circuit. When properly applied this results in: (a) a reduction in hum, (b) a reduction in noise, (c) a reduction in harmonic distortion, (d) an improvement in frequency response, and (e), in certain cases a reduction in the effective plate impedance of the output tubes, resulting in improved fidelity. These benefits apply only to those portions of the amplifier covered by the feedback network, which usually consists of a single resistor and/or condenser. This is connected to feed a portion of the output voltage back to the input circuit in phase.

In circuits where the feedback voltage is taken from the secondary of the output transformer, it is important to get the correct polarity of this winding. If incorrect, oscillations (a howl) will usually be set up or an increase of signal will result instead of a normal decrease. In this case it is only necessary to reverse the secondary winding for correct polarity. For example, if the feedback resistor connects to the 500 ohm tap on the output transformer and the common tap is grounded, merely reverse the connections. Now the feedback resistor connects to the common tap and the 500 ohm tap is grounded. This, of course, does not change the impedance taps for speaker matching in any way. As mentioned, connecting the inverse feedback network to an amplifier results in a loss of gain. The amount of loss is a measure of the amount of feedback inserted. Generally, about 20% is sufficient for good results. This loss may be compensated for by using another stage to increase the gain to normal.

Some forms of tone control are at times used in conjunction with the feedback network to increase or decrease the gain at certain frequencies. This type of control has many merits if not used for excessive compensation, for here we may introduce phase distortion and the amplifier may become unstable.

V. Motorboating.

Whenever you hear sounds like the exhaust of a motorboat, it is wise to look for the cause in the power supply or associated circuits. Motorboating is due to a common "B" supply impedance between two or more stages in an amplifier. The cure is fairly simple. Either increase the capacity of the last filter condenser or isolate the offending stage by a filter network which is generally called a decoupling filter. The latter usually consists of a resistor in the plate return lead and a condenser which by-passes this resistor to ground. In aggravated cases it is necessary to substitute a filter choke in place of the resistor. The latter should also be used in transformer coupled circuits as the choke will produce less voltage drop than a resistor.

VI. Parasitic oscillations.

When an amplifier overloads very easily and the tubes have short life, we may suspect that parasitic oscillations are occuring. These oscillations are usually of a frequency above the upper limits of hearing, so we must depend upon instruments to detect them--usually a low range A.C. voltmeter connected to the secondary of the output transformer. If an indication of voltage shows on the meter without feeding a signal to the amplifier, we may suspect parasitics. Any of the following remedies may be tried for the elimination of this difficulty:

a. Use of inverse feedback if not employed.

b. By-pass the plates of the output tubes with a .001 mfd or .002 mfd high voltage condenser to ground.

c. Put resistors of 100 to 1000 ohms in series with the grids of the output tubes.

d. Adequately by-pass all tube cathodes to ground.

e. Lower the screen grid voltage of the output tubes.

GENERAL CONSIDERATIONS

Although we have devoted our discussion to amplifiers alone, we must not forget its associated equipment. Whether it be a record player, a radio tuner, a microphone fed into the input stage, or the speaker connected into the output circuit, all are to be considered sources of distortion, just as defects in the amplifier itself. We may have the best quality amplifier obtainable, but if the associated equipment is of poor grade, corresponding results may be expected. Conversely, if we use high grade associated equipment, to do justice to these we should use a good amplifier.

Measurement of Frequency Response

Should a variable audio oscillator be available, we can

readily check the frequency response of the amplifier. Simply feed the output of the oscillator into a low range A.C. voltmeter, (or DB meter) and vary the oscillator frequency from one extreme to the other, recording at regular intervals the voltage indicated. Then disconnect the voltmeter from the oscillator and connect it to the voice coil terminals of the amplifier output transformer, thus connecting the oscillator to the input of the amplifier. Now the volume control on the amplifier is adjusted to give the same reading on the voltmeter as was obtained on the previous setup with the oscillator frequency set to 1000 cycles. Again, we vary the oscillator frequency from the two extremes as before, and note the new voltmeter readings, comparing them with the previous records for similarity. The differences in voltage may be translated into DB's by means of a voltage ratio to DB chart in order to have an accurate picture of the fidelity of the amplifier.

If no measuring instruments are available and you need a standard of comparison to judge your amplifier's response, it is suggested you listen to the sound in a good theatre and rate your audio system accordingly. If the comparison is favorable you will know there is little to be desired.

Proper Phasing of Speakers

When two or more speakers are connected to an amplifier and are located within the same room, the instantaneous excursions of the speaker cones should all be in the same direction. The speakers are then said to be in phase. Out of phase operation results in distortion and loss of volume. A good method of properly phasing speakers is to connect the voice coil of one to a $1\frac{1}{2}$ volt dry cell and watch the movement of the cone at the instant of contact. Then connect the dry cell to the other speaker in such a way that its cone also moves in the same direction as the first speaker at the instant of contact. Any number of speakers can be phased in the same manner. It is now only necessary to connect the various speakers, observing the voice coil polarity as determined, in order to have correct speaker phasing.

Using the Amplifier for Recording

Practically any type of amplifier can be used as a recorder, provided the output of the amplifier has sufficient capacity to actuate the recording head. There are two types of heads used for cutting acetate and nitrate cellulose recording blanks--magnetic and crystal cutters. Since both operate at different impedances, we must properly match the one we intend to use to the output of the amplifier. The magnetic cutter is of the low impedance type--generally around 8 ohms impedance; while the crystal cutter is a high impedance type, more nearly matching the plate impedance of the output tube. An undistorted output of 3 watts or more is required to drive the cutting head, depending upon type and make. A modulation indicating device is practically a necessity for recording in order to prevent over cutting of the record grooves. This indicator may be a neon lamp, a magic-eye tube or an output meter. In all cases the output level of the amplifier is adjusted so that the loudest signal does not produce an overcut. The indicator is continually watched while cutting the record and if overmodulation, i.e., if overcutting shows at any time, reduce the gain control to just below the point that overcutting takes place.

Connecting a Microphone to the Amplifier

There are four types of microphones in common use today: Carbon, Crystal, Dynamic and Velocity microphones. All except the crystal are essentially low impedance devices, and to connect them to a high impedance source, a transformer is used. This transformer is either contained in the microphone case or in the amplifier unit. Where, in these other three types of microphones, the transformer is external to the microphone case, long lines may be run between the microphone and the amplifier with very little loss. If a high impedance matching transformer is contained in the microphone case, only about 25 or 30 feet of cable can be used without affecting the fidelity of the entire unit. Long lines connecting a crystal to the microphone results in the loss of gain.

The carbon microphone is the most sensitive type of all; its voltage output, especially at voice frequencies, is the greatest. Nevertheless, it has the disadvantage of being noisy in operation and has the poorest frequency response curve. It also requires an external source of "button" current to operate correctly--something which is not convenient for the ordinary requirements of public address work.

The velocity microphone is the least sensitive of all types, but has the best frequency response. In using a velocity microphone, care must be taken not to speak directly into it; a minimum distance of 8 inches should be maintained between the speaker and microphone.

The crystal microphone has good frequency response, but because of its fragility, caution must be exercised in its handling. Care must also be taken with regard to extremely high temperatures and humidity.

The dynamic microphone is the sturdiest type of all, and if well constructed, has the advantage of ruggedness combined with good response. It is probably the best all around microphone for P.A. work.

Another type of microphone used, especially in intercommunication systems, is the ordinary permanent-magnet speaker. By means of switching, a speaker can be connected into the input of an amplifier, and by talking into the speaker the equivalent of a dynamic type microphone is obtained.

For intercom work this microphone arrangement is quite satisfactory as well as convenient.

In all cases, when connecting the microphone to the amplifier, it is advisable to use low capacity shielded cable; otherwise hum may be picked up in the line or the frequency response may be poor.

Acoustic Feedback

When a microphone is placed too close to the speakers, a

howl will be set up as the gain of the amplifier is increased. To prevent this feedback, careful thought to the placement of this equipment is required--and in severe cases, a good deal of experimentation! The better types of microphones will give a minimum amount of trouble of this kind, especially if the uni-directional varieties are used. High quality amplifiers and speakers are also of great help. Proper draping of the walls with a sound absorbent material to prevent sound reflections are often used when other techniques fail. Always try to keep the distance between the microphones and speakers as great as possible and place each unit so that the direct output of the speakers is not in line with the microphones.

Size of Amplifiers and Speakers

When there is a choice in making a public address installation, using the largest speakers and baffles available, and choosing the amplifier with the greatest power output, will always give the best results. The greater the power being drawn from a given amplifier and speaker, the greater will be the distortion. Even if the amplifier and speaker are capable of handling 25 watts, the distortion will be less at 10 watts output than at 20 watts, and lower at 20 watts than at 25 watts. The larger the diameter of the speaker and the larger the size of the magnet in permanent magnet types (which are used almost exclusively in P.A. work) or in the case of the field type speaker, the greater the amount of copper wire wound on the field, the more efficient will be the speaker. If we use a 25 watt amplifier to feed two speakers, each speaker will have to dissipate 12.5 watts at full, undistorted output. It is therefore necessary to know the power handling capacity of the speaker we intend to use. Economizing in this respect by using smaller speakers will only result in short life and excessive distortion. Most speaker manufacturers give the wattage ratings of their speakers in their catalogues; this information will help in determining the type of speaker needed. The baffle used with a speaker will determine the acoustical output available from a speaker. The horn type of baffle is most efficient but is too bulky for average use. A good bass-reflex baffle is probably the best for indoor use, or if a large flat surface can be utilized, such as a speaker built into a wall, the results will generally be good. The need for a large baffle will be apparent if we remember that the larger the effective area of the baffle, the better are the bass tones reproduced; also in this respect, the larger the diameter of the cone of a speaker, the more efficiently will the bass notes be reproduced.

Since the above discussion of amplifiers is limited in scope, it is therefore intended as a practical rather than a theoretical aid to the constructor and technician. If more theoretical information is required, it is suggested that the reader refer to any of the many good texts on the subject.

9 781882 580255

Made in the USA
Monee, IL
08 November 2026

67968603R00063